General Biology

Laboratory Studies in Biology

Second Edition

Paul W. Sattler
Robert O. Littlejohn
Terry L. Weaver
Steve W. Perry
Norman Reichenbach
Eugene D. Sattler
David A. DeWitt
Charles Detwiler

KENDALL/HUNT PUBLISHING COMPANY
4050 Westmark Drive Dubuque, Iowa 52002

This edition has been printed directly from camera-ready copy.

ISBN 10: 0-7575-3434-1
ISBN 13: 978-0-7575-3434-8

Printed in the United States of America
10 9 8 7 6 5 4 3

Contents

The Microscope and Basic Cell Structure

Introduction

In modern times, advances in our understanding of the world of science have usually followed advancements in scientific technology and instrumentation. For instance, the discoveries concerning the nature and consistency of the rings surrounding the planet Saturn were possible only because of the wonderful advancements that have been made in the areas of transportation and communication. Many similar examples of technology preceding discovery exist in the history of science.

With the development of the first microscopes in the mid 1600's a whole new world of discovery was opened to the biologist. Today, with the aid of compound light microscopes such as the one you will use in this lab course, humans are able to observe subjects up to 1000 times smaller than can be seen with the unaided eye.

Objectives

Upon completing your study of this unit, you should be able to:

1. Identify the major parts of the microscope listed in the illustration on the next page, and the functions of those parts.
2. Predict what the orientation of an object will be when viewed through the microscope by its orientation when viewed with the unaided eye.
3. Calculate the total magnification of the subject viewed when using any of the objectives on your microscope.
4. Calculate the diameter of the field of view when using high power from diameter and magnification values given for scan or low power objectives.
5. List the parts of an epithelial cell and an *Elodea* cell which are detectable with the compound light microscope.
6. Estimate the size of a cell by comparison with the diameter of the microscope's field of view when using a given magnification.

Exercise 1. Viewing Prepared Microscope Slides.

Note: The microscope you will be using is a very expensive, precision instrument. All instructions regarding its care and handling *must be strictly followed!* Two hands are necessary to properly carry a microscope. Place one hand on the ***arm*** at the rear of the scope (closest to you when you approach the cabinet) and the other hand under the ***base*** of the scope. Hold the scope upright at all times (some scopes have detachable parts which may fall out and break if not held upright).

Go to the microscope cabinet and carefully carry a scope to your lab table. Each student should have a scope for his or her personal use. Plug the scope in and turn on the

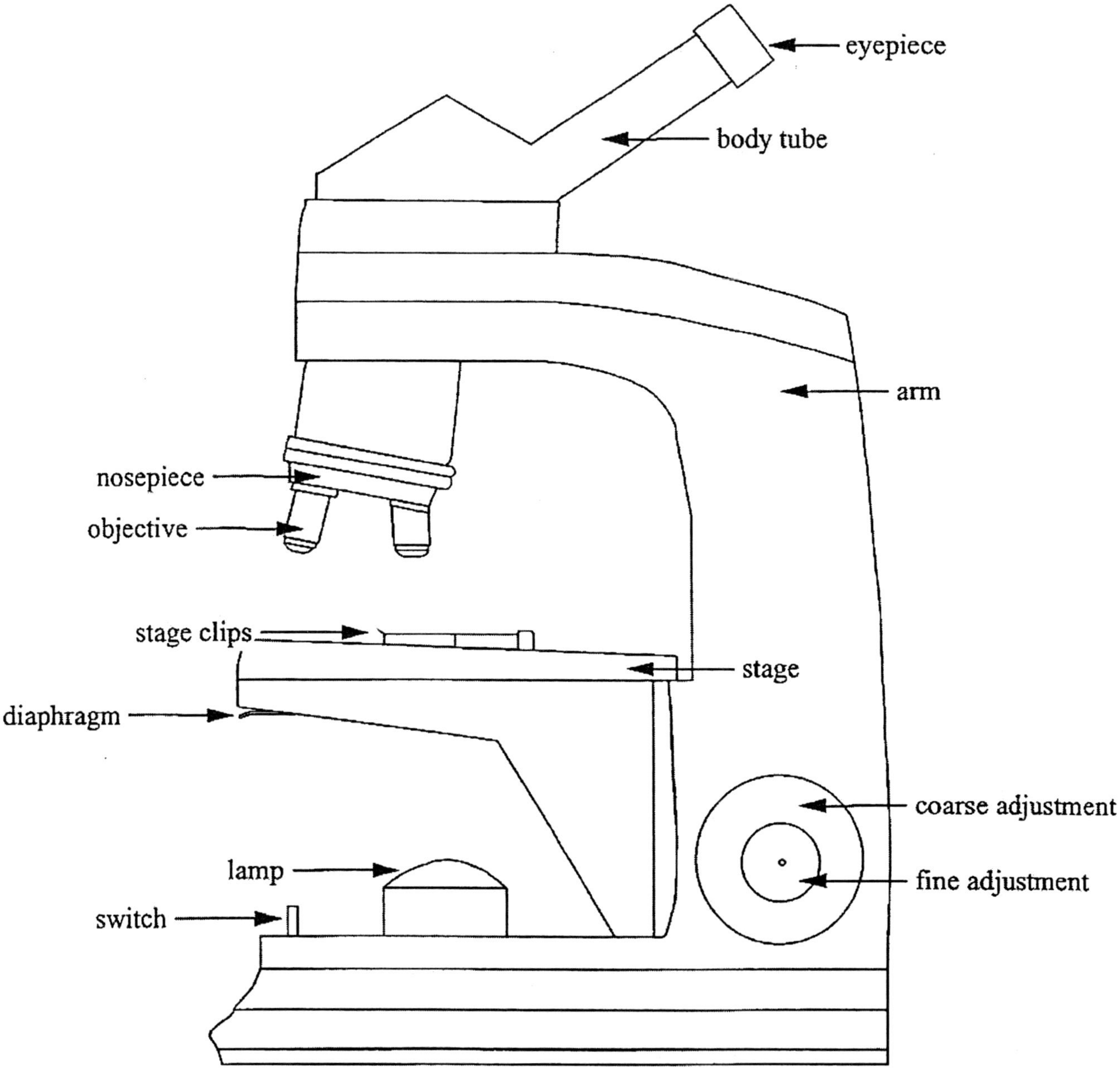
eyepiece
body tube
arm
nosepiece
objective
stage clips
stage
diaphragm
coarse adjustment
fine adjustment
lamp
switch

illuminator. To prevent accidents, it may be best to place the excess electrical cord in the book bin or drawer in your lab desk. Use the illustration on the opposite page to familiarize yourself with the parts of your microscope. Obtain a ***prepared slide*** of the letter "e" and a metric ruler for use in the exercise. (A prepared microscope slide displays an object which has been permanently mounted for repeated use in study.) Rotate the revolving ***nose piece*** so that the ***scanning objective*** (shortest objective, stamped "4X") is in the viewing position (in line with the ***body tube***). Be sure that the lens "clicks" into place. Place the "e" slide on the stage so that the letter "e" which is mounted under the glass cover slip is positioned directly in the center of the hole in the ***stage***. Secure the slide with the ***stage clips***. While looking at the stage assembly from the side, use the ***coarse adjustment*** knob to lower the objective to its lowest position. Focus the image by gradually raising the objective with the coarse adjustment while observing through the ***eyepiece***. (If focusing is impossible you may need to clean the eyepiece or objective. Ask your instructor for the special lens paper which will not scratch the lens.) The focus may be sharpened using the ***fine adjustment*** knob. By adjusting the size of the ***diaphragm*** beneath the stage you can adjust the contrast of the image against the light background, much as the contrast can be adjusted on a television. Experiment a bit to find the diaphragm position which gives the best contrast.

● a) In the space provided on your answer sheet, draw the mounted letter "e" as viewed through the microscope. Is the "e" backwards, upside down, right side up, etc.? ●b) Next draw it as viewed directly (look around the microscope arm directly at the slide). ●c) Besides a change in size, how is the orientation of the subject altered when viewed through the microscope? ●d) Considering that you are viewing the "e" through a 4X objective (which magnifies the object being viewed four times) compounded with a 10X eyepiece (which magnifies the objective image ten times), multiply the magnification of the two lenses to determine the total magnification. ●e) How many times larger does the "e" appear when viewed through the microscope using scanning power (4X) than it actually is? (Hint: this is an *easy* question!)

Next remove the "e" slide and place the metric ruler across the viewing area. Use the fine adjustment to bring the black graduation marks into focus. Using the millimeter (smallest) graduations, position the ruler so that the left edge of one of the black lines is at the left edge of the round, lighted viewing area. From the left edge of that black line to the left edge of the next black line is approximately one millimeter (mm). ●f) Estimate, to the nearest 0.1 mm, the diameter of the field of view on scan power. Record the diameter on your answer sheet.

Remove the ruler and replace the "e" slide on the scope. Adjust the focus once again. Carefully rotate the nosepiece (while watching from the side) so that the low power objective (10X) is in the viewing position. (Rotate the nosepiece directly to the 10X objective rather than passing by the high power first. This will prevent damage to the slide and the objective.) If your microscope is ***parfocal***, the "e" will be in rough focus when using either the low power (10X) or high power (43X) objectives, once you have focused using the scan (4X). Use the fine adjustment to sharpen the image. You will likely need to adjust the diaphragm again because changing the objective changes the quantity of light which enters the scope.

●g) Calculate and record the total magnification on low power, using the formula on the answer sheet. Remove the "e" slide and use the ruler as before to estimate, to the nearest 0.1 mm, the diameter of the field on low power. ●h) Record the diameter on your answer sheet. ●i) Is the diameter using 10X larger or smaller than the diameter for the scanning objective? ●j) Offer an explanation for this.

Remove the ruler and replace the "e" slide once more. Adjust the focus and rotate the nose piece (while observing from the side) so that the high power objective (43X) is in the viewing position. (**Note:** never use the coarse adjustment while the high power objective is in position.) Sharpen the focus using the fine adjustment knob and adjust the diaphragm for best contrast. ●k) Can you see more or less of the "e" than before? ●l) Offer an explanation for this. ●m) Calculate and record the total magnification on high power.

Because of the high magnification you will not be able to use the ruler to estimate the field of view. Instead, use the following equation and your previous measurements to estimate the diameter of the field of view on high power:

$$\text{Dia (high power)} = \frac{\text{Dia (low mag) x Mag. (low mag)}}{\text{Mag (high mag)}}$$

in which Dlow and Mlow are the Diameter (which you measured with the ruler) and Magnification (which you calculated) of the *low* power objective, and Dhigh and Mhigh are the unknown Diameter and the known Magnification of the *high* power objective (from Question **m**). ●n) Record the diameter on your answer sheet.

Note: Your microscope may be equipped with a fourth objective marked 100X. We will not be using it in this laboratory. *Do not rotate this lens into the viewing position!* To do so may damage the lens.

The Cell

One of the first discoveries made using a microscope was that of the existence of cells by Robert Hooke in 1665. Hooke noticed the highly ordered arrangement of tiny compartments in a slice of cork. He named the compartments "cells" because they reminded him of little rooms such as those found in a prison. From studies which followed this discovery, the cell theory arose which states as one of its axioms that all living things are composed of cells.

Exercise 2. Preparing and Viewing Human Epithelial Cells.

Obtain a clean microscope slide and cover slip. Place one drop of 0.01% methylene blue stain in the center of the slide, and set it carefully on your lab table. **Caution:** Do not get the stain on your clothes… it will not come out! Using a clean toothpick, *gently* scrape the inside of your cheek a number of times and stir or rub the same side of the toothpick around in the stain on your slide.

Hold the cover slip by its edges, allowing the third edge to rest on the slide. Drag the third edge of the cover slip across the slide toward the stain. When the stain touches the cover slip it will spread out along the bottom edge. Slowly allow the slip to drop down over the stain. This method will push any air bubbles which might have formed out from under the cover slip. You have just prepared a ***wet mount***.

Place your wet mount on the microscope stage just like you did with the "e" slide and proceed exactly as before, beginning with scanning power. While using low power, locate a few cells (they're very tiny!) and see if you can distinguish some cell parts. Don't forget to adjust the diaphragm for best contrast.

The outer edge of the cell is defined by the presence of a ***cell membrane***. You won't actually be able to see the membrane, but you should be able to locate it by determining where the ***cytoplasm*** stops. The cytoplasm is the clear liquid solution in the cell. It is within this solution that many of the metabolic processes of the cell take place. The membrane primarily regulates the exchange of materials between the cytoplasm and the cell's environment.

You should be able to see a dark dot in the center of the cell. This is the ***nucleus*** which houses the hereditary information of the cell. Sketch a few of your epithelial cells in the space provided on your answer sheet, and label the three basic parts.

Change to high power and estimate the diameter of the epithelial cells by judging the number of cells that could fit edge-to-edge across the field of view and dividing that number into the diameter of the field.

$$\text{Cell diameter} = \frac{\text{Diameter of field at current magnification}}{\text{Number of cells which could fit across field of view}}$$

●a) Record the cell diameter on your answer sheet. ●b) Give your answer in micrometers (μm) as well as in millimeters (1 μm = 1000 mm). To do this, multiply your answer in millimeters times 1000.

Exercise 3. Preparation and Viewing of Plant Cells.

Obtain a single leaf of *Elodea* and make a wet mount using water instead of methylene blue. You may need two drops of water because of the thickness of the leaf. **Note:** Extra care should be taken (especially on high power) not to break the cover slip.

Observe the leaf cells on scan and then on low power. Adjust the contrast. You will find the leaf to be more than one cell layer thick. Once you have the leaf in rough focus, use the fine adjustment to focus up and down through the different cell layers. This will give you a feel for the depth of focus of your scope.

●a) How many layers of cells can you detect in the leaf? Now estimate the length of the cells by judging how many will fit lengthwise across the field of view. ●b) Approximately how many millimeters (mm) long are the cells? ●c) How many micrometers (mm) long?

Like animal cells, plant cells have cell membranes, cytoplasm and nuclei. Plant cell membranes, like animal cell membranes, are not visible with the compound light microscope. However, plant cells have ***cell walls*** which are located just outside the membrane. Cell walls are clearly visible with the light microscope and appear as rigid structures which separate the individual cells in plant tissue. Plant cells also have a ***central vacuole*** (a membrane-bound sac within the cell membrane) which takes up most of the volume of the cell. Since membranes are not visible, the vacuole will appear as an empty circle in the cell center. The vacuole serves as a storage area for metabolic compounds and waste products. Between the vacuole and cell membrane are lots of green ***chloroplasts***. These ***organelles*** (the chloroplasts) are the sites of photosynthesis.

Using high power, try to locate each of the cell parts mentioned. Remember, membranes are not visible using the light microscope, but you can determine the location of the vacuole by the arrangement of the chloroplasts around it. You should also know that unless a stain is used, you will not see the nucleus in the plant cell. Sketch a few *Elodea* cells in the space provided on your answer sheet and label the parts observed. ●d) What structures did you observe in the plant cells which were not present in the epithelial cells?

As the leaf cells begin warming in the microscope's light source, ***cyclosis*** (cytoplasmic streaming) may begin to occur. You may notice the movement of the chloroplasts about the vacuole as they are swept along in the flow of the cytoplasm. Observe this process for a few minutes and see if you can detect a pattern in the direction of flow.

When you are finished with your wet mounts, place them in the pan of cleaning solution provided in the laboratory. Use the same care in returning your microscope to the cabinet that you did when you got it (assuming you were careful then). Before returning your scope to its cabinet:

1. Remove slides from the stage.
2. Clean and dry the stage and all objectives with lens paper.
3. Place the scanning objective in the viewing position.
4. Turn the coarse adjustment knob all the way forward.
5. Turn the fine adjustment knob all the way forward, then back two full turns.
6. Loosely wrap the cord around your hand and set it over the lamp.

Name ____________________________

Number __________________________

The Microscope and Basic Cell Structure

Exercise l. Viewing Prepared Microscope Slides.

a. ____________________ b. _______________

c.

____

d. 4X x l0X = total magnification on scan = ______ X

e. _____ times

f. _____ mm

g. l0X x l0X = total magnification on low = _______ X

h. _____ mm

i. _________________________

j. Explanation:

k. _________________________

l. Explanation:

m. 43X x l0X = total magnification on high = ________ X

n. _____ mm

Exercise 2. Preparing and Viewing Human Epithelial Cells.

Sketch epithelial cells below:

a. _____ mm b. _____ µm

Exercise 3. Preparation and Viewing of Plant Cells.

a. __________ layers b. __________ mm c. _____ µm

Sketch *Elodea* cells below:

d.

____

Diffusion and Osmosis

Introduction

The function of the cell membrane is to separate the cell from its "hostile" environment. It is this separation that makes life possible. How then do raw materials get into these "separated" cells, and how do waste products get out? How do molecules move around inside a cell once they get there? These are the concepts that will be examined in this lab.

Cells basically function in an aqueous (water) environment. Water is the ***solvent*** (the material in which other materials dissolve) of life. The chemicals important to a cell are dissolved in the water and therefore are ***solutes*** (materials that are dissolved in solvents). The atoms and molecules of the solutes mix with those of the solvent to form a ***solution*** and must be considered together to understand how things get into and out of cells.

Objectives

Upon completing your study of this unit, you should be able to:

1. Define solute, solvent, solution, diffusion, and osmosis.
2. Describe Brownian movement.
3. Describe the effect of concentration on rate of diffusion.
4. Describe the effect of molecular weight on rate of diffusion.
5. Define tonicity, hypotonic, hypertonic, and isotonic.
6. Given solute concentrations inside and outside a cell or the tonicity of a cell's environment, be able to predict whether water will move into or out of a cell.
7. Define turgidity, plasmolysis, and hemolysis.
8. Explain the function of the central vacuole of plant cells and the role it plays in turgidity.
9. Describe the effect of distilled water on red blood cells.

Note: Because of the time required for Exercise 4, set it up at this time. Then set up Exercises 2 and 3. Return to this point and work through the lab.

Diffusion

Diffusion is the movement of atoms or molecules of a substance from high to low concentration. Such movement is said to be in the direction of the concentration gradient or "with" the concentration gradient (high concentration to low concentration). But just how does a molecule move?

The solution to this question is found in the 19th century observations of Robert Brown. He observed that tiny particles in water when viewed under the microscope appeared to be vibrating, bouncing, and moving about. Eventually Brown correctly suggested that the movement was due to bombardment of the particles by molecules. We now know that water molecules hitting the particles cause the phenomenon observed by Brown. The constant molecular motion characteristic of molecules has been called ***Brownian motion*** and is driven by heat energy in the environment.

Exercise 1. Brownian Motion.

Place a drop of water containing a stain on a microscope slide. Place a cover slip on the drop. Focus using scan power, then low power, and finally on high power. Adjust the diaphragm for the clearest view, and observe the movement of the tiny stain particles. ●a) Describe the Brownian motion on your answer sheet. ●b) Is the movement of an individual particle directional (a straight line) or random (vibrating, oscillating, and changing directions)?

Brownian (random) movement of molecules causes them to spread out and move from an area of high concentration to areas of lower concentration, as they ricochet off one another. This is ***diffusion***. In the next two exercises you will examine some of the important factors influencing diffusion.

Exercise 2. Effect of Concentration on Diffusion.

Methyl red is a pH indicator that changes from yellow to red as the solution changes from basic to acidic. Obtain a pre-poured agar plate and use a soda straw to punch three holes in a triangle well-separated from each other and from the edge of the dish.

Carefully remove the agar plugs with a dissecting needle. Turn the dish over and label the wells as #1 - #3 by writing on the bottom of the dish with a marking pencil.

Fill well #1 with 1.0M HCl (hydrochloric acid). Be sure to get the correct molarity. Molarity (M) is a unit of concentration where higher molar values represent greater concentration. Count the drops required to fill the well just to the top, being careful that you don't cause the well to overflow. Put in well #2 the same number of drops of 0.1M HCl and in well #3 an equal number of drops of 0.01M HCl.

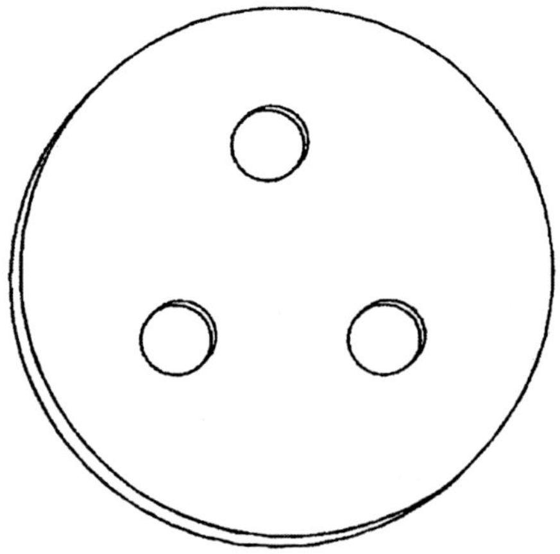

Allow the plate to sit undisturbed while the HCl diffuses out into the agar. After a half hour measure the diameter of the diffusion zone with a millimeter ruler. A sheet of white paper under the dish will make the measurements easier. ●a) Record your data on the answer sheet.

Important Principle: When graphing scientific data, there are usually two quantities involved: one that the scientist controls and one that he or she observes. The quantity being controlled is usually plotted on the X (horizontal) axis, and the quantity being observed is plotted on the Y (vertical) axis. In this exercise we *controlled* the acid concentration, and we *observed* the diffusion distance.

●b) Graph the acid concentration on the X axis and the distance of diffusion in mm on the Y axis. Select units for the diameter that will use a maximum area of the graph. ●c) What is the effect of concentration on diffusion?

Exercise 3. Effect of Molecular Weight on Diffusion.

Prepare another methyl red agar plate as in the previous exercise. Label the wells #1– #3. Add an equal number of drops to each well as follows:

#1 – Formic acid (molecular weight= 46)
#2 – Malic acid (molecular weight = 134)
#3 – Citric acid (molecular weight = 192)

Leave the plates undisturbed while the acids diffuse out into the agar. Near the end of the lab period measure the diameter of the diffusion zone with a millimeter ruler. ●a) Record your data on the answer sheet. ●b) Which quantity (molecular weight or diffusion distance) did you intentionally vary by selecting different acids? •c) Which quantity did you observe changing during the exercise? ●d) Based on the "important principle" at the top of the page, which quantity should you graph on the X axis? ●e) Which on the Y? ●f) Graph your results, remembering to use units that will use as much as possible of the graph area. ●g) What is the effect of molecular weight (size) on diffusion?

Note: Dispose of the agar plates in a trash can.

Osmosis

You have seen that molecules move by diffusion from high to low concentration. There are three very important molecules that readily enter and/or exit cells by diffusing through the cell membrane: O_2 (oxygen), CO_2 (carbon dioxide), and H_2O (water). Most other important molecules cannot pass freely through the membrane. However, once inside the cell they move by diffusion. The passage of these other molecules is regulated by proteins in the membrane through a variety of mechanisms collectively called transport processes. Thus with the exception of O_2, CO_2, and H_2O the cell controls what gets in and what gets out by which transport proteins are made.

Because water is the solvent of life and because it freely diffuses through cell membranes, there are some special considerations related to the movement of water. The diffusion of water into and out of cells is called ***osmosis***. Water diffusion is like all other diffusion in that water molecules move from high to low concentration. However, since water is the solvent, we are more used to thinking about solute concentrations. To keep this straight one has only to remember that since two molecules cannot occupy the same space, the higher the solute concentration the lower the water concentration will be. With regard to osmosis, the following three situations are possible:

Tonicity of Solution:	**1.** Hypertonic	**2.** Hypotonic	**3.** Isotonic
Solute Concentration :	Higher outside cell than in	Lower outside cell than in	Same inside and outside
Water Concentration :	Higher inside cell than out	Lower inside cell than out	Same inside and outside
Water moves:	Out of cell	Into cell	Stays same

In situation #1 the tonicity (measure of solute concentration outside the cell relative to inside) is high which means water concentration is greater inside the cell; therefore, water diffuses out of the cell. Situation #2 is just the opposite. Low tonicity means that water concentration is higher outside the cell and water molecules diffuse into the cell. The last possibility, isotonic, describes the situation when water concentrations are the same inside and outside the cell, and there is no net diffusion in either direction. Animal cells are stable in an isotonic solution, but tend to have problems in hypertonic and hypotonic environments.

Exercise 4. Osmosis with Simulated Cells.

In this study you will construct simulated cells and observe osmosis. The key to simulating a cell is to have a membrane that will permit water, oxygen, and other small molecules to pass through but which will not allow the passage of larger molecules. Cellophane is such a material with tiny pores which permit water to move freely through by diffusion yet which retain larger molecules.

Soak three 2.5" pieces of cellophane tubing in water until flexible. Rub the tubing between your fingers to open it. Then tie a piece of string tightly around one end. Trim short the ends of the string. Using a dropper, fill each of the three tubes about 3/4 full with 25% corn syrup. Then tie a piece of string around the open end to form a tightly closed bag.

Trim short the ends of the string. The cellophane simulates the cell membrane and the corn syrup the cytoplasm of a cell. Corn syrup, like cytoplasm, is mostly water in which a variety of larger solute molecules are dissolved. These molecules are too large to pass freely through the membrane.

Place 50ml of the three following syrup concentrations in separate small beakers. Label the beakers #1 - #3:

#1 – 50% syrup - hypertonic
#2 – 25% syrup - isotonic
#3 – 0% syrup - hypotonic

●a) Dip each simulated cell in water, blot on paper towel, and record its weight (nearest 0.1 gram) on your answer sheet. Place one cell in each beaker. Allow the "cells" to sit for a full hour. While you are waiting, discuss what should happen, based on how osmosis works. ●b) Record your prediction on the answer sheet.

●c) After one hour, rinse each cell in water, blot on paper towel, and record its weight on your answer sheet. Subtract the final weights from the initial weights to determine weight change. Use "+" to indicate a gain in weight and "-" to show a weight loss. Were your predictions correct? ●d) Describe the appearance of each "cell" and write your conclusion about the tonicity of each beaker. ●e) Explain the changes in appearance and weight of each simulated cell in respect to osmosis.

Now that you've had the opportunity to study osmosis with simulated cells, we'll examine some live cells and their response to changes in tonicity. Not all cells respond the same way to environmental changes because cells are physiologically and structurally different. Plant cells, prokaryotes, fungi, and algae all have rigid cell walls that will permit the uptake of only a limited amount of water. When these cells are in a hypotonic environment, water uptake pushes the cell membrane out tightly against the cell wall creating a pressurized taut condition referred to as ***turgidity***. Depletion of water in plants can be observed as the lack of turgidity referred to as wilting.

Exercise 5. Plant Cell Turgidity.

As a demonstration, two thin slices of potato have been cut and placed, one in a beaker containing distilled water (no solute), and the other in 3% NaCl (sodium chloride or table salt). Allow osmosis to proceed for an hour or so, then feel each slice. Try to bend the potato slices. ●a) Record your observations on the answer sheet. ●b) If you wanted to serve a crisp lettuce salad, would it be better to soak the lettuce in distilled water or salt water?

Exercise 6. Microscopic Observation of Osmosis in Plant Cells.

Tear off an *Elodea* leaf, place on a slide in a drop of water. Add a cover slip and examine under the microscope using the low power objective. Then switch to the high power objective. You should see the typical box-shaped plant cells with their green chloroplasts. The chloroplasts tend to be located near the edge of the cell because the central water vacuole is full, and it pushes the cytoplasm with its chloroplasts, nucleus, and other organelles to the edge of the cell. ●a) Draw *Elodea* on your answer sheet, and indicate the location of the central vacuole, chloroplasts, cell membrane, and cell wall.

After recording your initial observations on the answer sheet, place a drop of 3% NaCl solution on the edge of the cover slip. By touching a piece of paper towel to the other side of the slip, draw the NaCl under the slip. Alternatively, you can remove the cover glass, add the NaCl, and replace the cover glass. Observe the *Elodea* cells over the next few minutes. Shrinkage of a plant cell away from the cell wall is called ***plasmolysis***. ●b) Draw the *Elodea* as it now appears, and again indicate the location of the central vacuole, chloroplasts, cell membrane, and cell wall. ●c) Explain what happened.

Exercise 7. Observation of Osmosis in Animal Cells.

Animal cells differ greatly in their response to osmotic imbalance because they lack the rigid cell wall that prevents plant cells from swelling up and bursting. ●a) What would happen to an animal cell in a hypotonic environment?

Number three tubes #1-#3 and add the following:
Tube #1 – 5.0 ml tap water
Tube #2 – 2.0 ml tap water + 3.0 ml isotonic saline (0.9% NaCl)
Tube #3 – 5.0 ml isotonic saline

Note: Before continuing, be sure to read the following paragraph all the way through so that you won't miss what happens.

Add a drop or two of blood to each tube and mix by shaking gently. Watch to see if an *immediate* change takes place. Red blood cells take up water and burst (***hemolysis***) in a hypotonic environment. You can observe that hemolysis has occurred by a clearing in the blood cell suspension that enables you to see the type on this page through the tube. ●b) Record the hemolysis time for each tube on your answer sheet. If hemolysis does not occur within one minute, record as no hemolysis. ●c) Explain your results.

Name ______________________________

Number ____________________________

Diffusion and Osmosis

Exercise l. Brownian Motion.

a. Description:

b.

Exercise 2. Effect of Concentration on Diffusion.

Well	Molar Concentration	Diameter of Diffusion Zone (mm)
1	1.0	__________
2	0.1	__________
3	0.01	__________

b. Graph concentration (X) vs. diffusion zone diameter (Y).

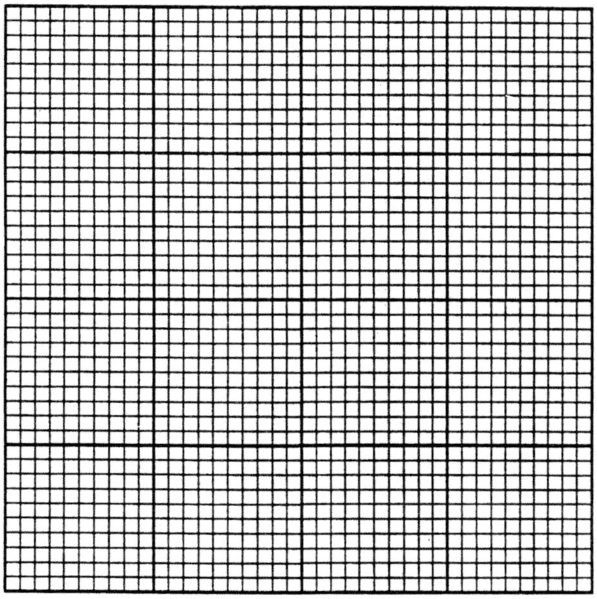

c. Effect:

Exercise 3. Molecular Weight and Diffusion.

a.

Well	Acid	Molecular Weight	Diameter of Diffusion Zone (mm)
1	Formic	46	__________
2	Malic	134	__________
3	Citric	192	__________

b. ______________________________

c. ______________________________

d. ______________________________

e. ______________________________

f. Complete the graph, and label the axes.

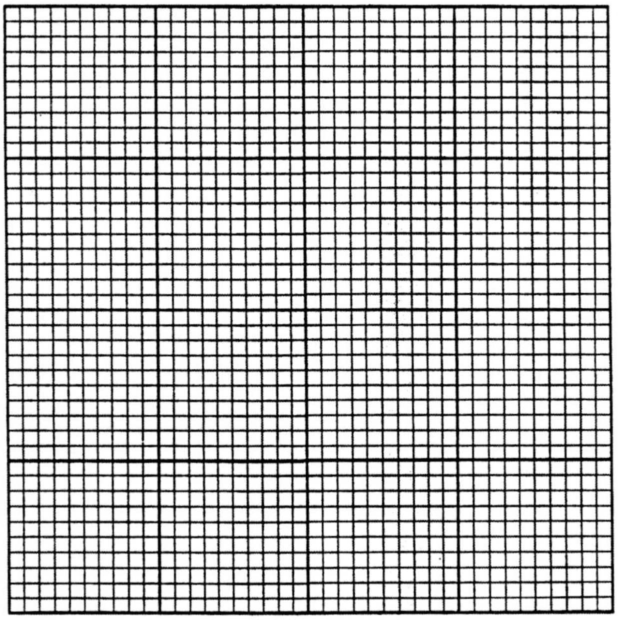

g. Effect:

Exercise 4. Osmosis with Simulated Cells.

	Concentration of Syrup in Beaker		
	0%	25%	50%
c. Final weight	________	________	________
a. Initial weight	________	________	________
	(Subtract initial weight from final weight)		
Change in weight	________	________	________
d. Appearance	________	________	________
Tonicity*	________	________	________

* (hypertonic, hypotonic, or isotonic—you can leave out the "tonic")

b. Prediction:

e. Explain your results:

Exercise 5. Plant Cell Turgidity.

a. Observation:

b. Which?

Exercise 6. Microscopic Observation of Osmosis in Plant Cells.

a. Drawing of cell:

b. Drawing of cell after plasmolysis:

c. Explanation:

Exercise 7. Observation of Osmosis in Animal Cells.

a. What?

b. Tube	Contents	Hemolysis Time
1	Water	________
2	Water & saline	________
3	Saline	________

c. Explanation:

Enzymes

Introduction

Each of the various processes we collectively call "metabolism" represents the composite of hundreds of individual chemical reactions. Under laboratory conditions, almost all of these reactions would proceed at such a slow rate that metabolism would be virtually impossible. Cells overcome this chemical deficiency by making specific catalysts that greatly speed up the rates of the individual reactions. These catalysts of living cells are called ***enzymes***. Cells regulate which chemical reactions will occur when by regulating which enzymes they make. The information needed to make a specific enzyme is in the genes (the DNA).

Enzymes are large proteins comprising hundreds of amino acids arranged in a particular unique sequence. It is the unique sequence of amino acids of an enzyme that confers its unique catalytic properties. The amino acid sequence results in a three-dimensional structure that is unique for each enzyme. Each enzyme has an ***active site*** which is the location where reactant (***substrate***) molecules attach and where the chemical reaction takes place. Anything that interacts with an enzyme causing a change in shape, and therefore a change in the active site, will inhibit enzyme activity. Chemicals which cause enzyme inhibition are usually referred to as poisons or toxins. When an enzyme's shape has been altered so that it no longer catalyzes its reaction, it is said to have been ***denatured***. In this study, we will examine some of the properties of enzymes and the important factors that influence them.

Objectives

Upon completing your study of this unit, you should be able to:

1. Define enzyme, active site, denaturation, denaturation point, buffer and substrate.
2. Write the Catalase chemical reaction.
3. Describe how to prepare a plant enzyme extract.
4. Describe the effect of enzyme concentration on enzyme activity.
5. Describe the effect of temperature on enzyme activity.
6. Describe the effect of pH on enzyme activity.
7. Name an inhibitor of Catalase.
8. Write a generalization for most enzymes regarding optimum pH and optimum temperature.

Enzyme Extraction and Preparation

In this study we will use an enzyme, Catalase, which is widely distributed in nature among many plants, animals, and microorganisms. Catalase catalyzes the following reaction:

Hydrogen peroxide (H_2O_2) is generated from a variety of necessary metabolic reactions in organisms, but is somewhat toxic and must be quickly removed from the cells. Perhaps you have used H_2O_2 as a disinfectant on a cut or scrape injury. If you have applied it to a wound, you may have noticed a lot of bubbling. That bubbling was the result of Catalase from your cells breaking down the H_2O_2 into water and oxygen.

In this part of the exercise, you are going to prepare an enzyme extract to use in the rest of the study. Potatoes or turnips are readily available and have relatively high levels of Catalase activity.

Exercise 1. Enzyme Extraction

Place 3 pieces of fresh potato or turnip along with 100 ml water in a blender. Blend for 10-20 seconds at high speed to break apart the plant cells and release the enzyme. If a blender is not available, a mortar and pestle may be used with smaller pieces of plant tissue and less water.

Pour the extract into a 100 ml beaker, and return to your lab station. Filter the broken cell/tissue extract through a single coffee paper filter into another 100 ml beaker. This should give you enough enzyme extract to use for the entire experiment.

Note: When measuring with a graduated cylinder, you will notice that the surface of the liquid is not flat but rather is lower in the middle. Always measure from the lowest part.

•a) Measure the amount of enzyme extract you have by pouring the extract into a 100 ml graduated cylinder. Record your results on the answer sheet. If you have less than 60 ml you will need to prepare more extract.

Exercise 2. The Catalase Reaction

Note: In order to avoid contamination of the hydrogen peroxide (H_2O_2) in the stock bottle, a waste beaker has been provided in which you should discard any excess. Do not pour anything into the stock bottle of H_2O_2.

Using a 10 ml graduated cylinder, add 5 ml of enzyme extract to a clean 20 ml test tube. Next, using a second 10 ml graduated cylinder, measure 5 ml of hydrogen peroxide (H_2O_2) and add to the enzyme extract. Reserve the graduated cylinders for use with extract and H_2O_2, respectively. Hold your thumb over the top, and invert the tube once.

Measure the height of the bubbles in the tube (from the top of the liquid to the top of the bubbles) using a metric ruler. .a) Record the height in millimeters (mm) every 30 seconds for 5 minutes or until the bubbles reach the top of the tube. The bubbles should reach at least 30 to 50 mm in two to five minutes and should not reach the top of the tube until two minutes have elapsed. If the bubbles don't reach 30 mm in five minutes, your enzyme solution is too dilute. You will have to discard your enzyme extract and prepare new enzyme extract, starting with twice as much plant material as you used the first time. If the bubbles reach the top of the tube in less than two minutes, the enzyme solution is too concentrated, and you will have to dilute it with water. To dilute the enzyme extract, mix it well with an equal volume of water (half and half). Repeat Exercise 2 with your new enzyme preparation. Continue with this process until you have an enzyme extract with the appropriate activity as described at the beginning of this paragraph. This is important for the success of succeeding exercises.

•b) Graph your results with time on the X (horizontal) axis and height on the Y (vertical) axis. Select the units for the graph so as to use a maximum amount of the graph surface.

In the following exercises you will be using the bubbles in different experiments as a way of measuring enzyme activity. •c) What is the reaction product that causes the bubbling? (Hint: look at the equation at the top of the previous page.)

Effect of Enzyme Concentration

Enzymes are not consumed in the reactions they catalyze. They bind ***substrates*** or reactants, catalyze the reaction, release the products, and proceed to repeat the process over and over again. Therefore the more enzyme molecules present, the more reaction will occur. If the gene which produces the enzyme is “turned off”, no enzyme is made; conversely, when a gene is “turned on” that particular enzyme is made. In this way cells regulate enzyme levels by controlling which genes are expressed at any given time and by regulating the number of enzyme molecules synthesized.

Exercise 3.

Label four tubes #1 – #4 with the marker provided, and place them in a test tube rack. Place the following quantities of enzyme and water in the tubes:

#1	0 ml enzyme	+	5 ml water
#2	1 ml enzyme	+	4 ml water
#3	3 ml enzyme	+	2 ml water
#4	5 ml enzyme	+	0 ml water

The tube with no enzyme is called an experimental ***control***. A control such as this should be included to verify that the effect is due to the variable in which we're interested, in this case, the enzyme concentration.

Place 5 ml H_2O_2 in each of four *empty* test tubes, bringing the number of test tubes in use to eight. When all eight tubes have been prepared, *quickly* add 1 tube of H_2O_2 (5 ml) to each of tubes #1 – #4. Immediately mix as described previously. Watch the enzyme activity. When the most active tube has produced bubbles extending well up into the tube (about 2-5 minutes), quickly measure the height of the bubbles in all four tubes. •a) Record your data on the answer sheet and graph your results. •b) Do higher or lower enzyme concentrations favor metabolic reactions? •c) How can the cell control the rate of a metabolic reaction? •d) Why did you include a tube with no enzyme?

Effect of pH on Enzyme Activity

The unit used to measure acidity is pH. Acidity is caused by hydrogen ions, and pH is a measure of hydrogen ion concentration. The pH scale extends from 0 to 14 with 7 being neutral, the acidity of distilled water. Very low pH values correspond to high hydrogen ion concentrations and therefore are highly acidic, while high pH values represent low hydrogen ion concentrations which are referred to as being alkaline or basic. Enzyme activity is affected by pH because changes in pH influence the shape of enzymes. Each enzyme has an *optimum* pH, a pH at which it is most active. Most enzymes have an optimum pH somewhere between 5.5 and 8.5 (around neutrality). This is consistent with the fact that cytoplasm tends to be around pH 7. Cells make molecules called buffers which regulate internal pH and protect against sudden pH changes.

Exercise 4.

Label three tubes #1 – #3, and using a clean 10 ml graduated cylinder for each solution, add the following:
1 — 4 ml enzyme + 1 ml pH 2.0 phosphate buffer,
#2 — 4 ml enzyme + 1 ml pH 7.0 phosphate buffer, and
#3 — 4 ml enzyme + 1 ml pH 12.0 phosphate buffer.

Place 5 ml H_2O_2 in each of three *empty* test tubes. When all three tubes are ready, *quickly* add the H_2O_2 to tubes #1-#3 and mix. When the bubble height in the most active tube has progressed well up into the tube (2 – 5 minutes) quickly measure the bubble height in all tubes. •a) Record your data on the answer sheet and graph your results. •b) Which tube was acidic? •c) Which tube was basic? •d) Which tube was neutral? Living cells of all types are distressed, irritated, or even killed by coming in contact with acidic or basic solutions because of the effect these solutions have on enzyme activity. •e) Which tube (pH) appears to be optimum for the enzyme? •f) Discuss what your results suggest with regard to the use of pH-balanced shampoos in which the pH is adjusted to about pH = 7. (Hint: enzymes are proteins, and so is hair.)

Effect of Temperature on Enzyme Activity

The general effect of an increase in temperature on chemical reactions is to increase their rate. However, proteins are relatively delicate biochemicals that denature if the temperature gets too high. Thus enzyme reactions speed up with temperature increases up to the denaturation temperature of the enzyme. At temperatures above the ***denaturation point***, reaction rates decrease due to destruction of the enzyme. Most enzymes begin to denature around 45°C.

Exercise 5.

Label three tubes #1 – #3, and place 5 ml enzyme extract in each tube. Into three *empty* tubes place 5 ml H_2O_2. Place tube #1 and one of the H_2O_2 tubes (not mixed, but separately) into a beaker of ice water (0°C). Leave tube #2 and one of the H_2O_2 tubes at room temperature. Place tube #3 and one of the H_2O_2 tubes (not mixed, but separately) into a boiling water bath (100°C). Wait 5 minutes for all temperatures to equilibrate, and then mix each H_2O_2 tube with its enzyme tube. Return each tube to its prescribed temperature. When the bubble height in the most active tube has progressed well up into the tube (2-5 minutes) quickly measure the bubble height in all tubes. •a) Record your data on the answer sheet and graph your results. •b) Which temperature treatment was optimum for enzyme activity? •c) Explain why organisms have a temperature range with a maximum and minimum outside of which they cannot survive.

Enzyme Inhibition

Chemicals in the vicinity of enzymes often influence protein shape and therefore influence enzyme activity. Changes in enzyme shape usually cause a decrease or inhibition of enzyme activity. Enzymes also have an optimal concentration of normal dissolved chemicals, and departure from that concentration can cause inhibition. Other chemicals cause inhibition even in very tiny concentrations. Heavy metals such as lead, mercury, and copper inhibit a broad range of enzymes and are harmful to living organisms.

Exercise 6.

Label two tubes #1 and #2, and add 4 ml enzyme to each tube. Add 1 ml water to tube #1 and 1 ml 20% copper sulfate to tube #2. To two *clean* tubes add 5 ml H_2O_2 and then mix one of these with tube #1 and the other with tube #2. •a) Compare the enzyme activity between the tubes and record your data on the answer sheet.

•b) Does copper affect enzyme activity? •c) Based on your observations, explain why heavy-metal-containing wastes and environmental "pollutants" threaten human health.

Name ______________________________

Number ____________________________

Enzymes

Exercise 1. Enzyme Extraction.

a. ____________________ ml.

Exercise 2. Enzyme Reaction.

a: Data:

Time (Seconds)	Height (mm)
0	____________
30	____________
60	____________
90	____________
120	____________
150	____________
180	____________
210	____________
240	____________
270	____________
300	____________

b. Graph: Time (X) vs. Bubble Height(Y)

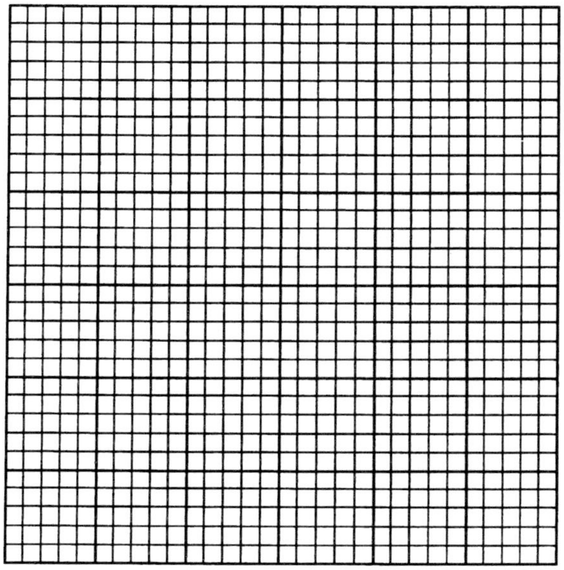

c.__________________________

Exercise 3. Effect of Enzyme Concentration.

a.

Tube	Quantity of Enzyme (ml)	Bubble Height (mm)
1	0	___________
2	1	___________
3	3	___________
4	5	___________

Graph: Enzyme Quantity (X) vs. Catalase Activity (Y)

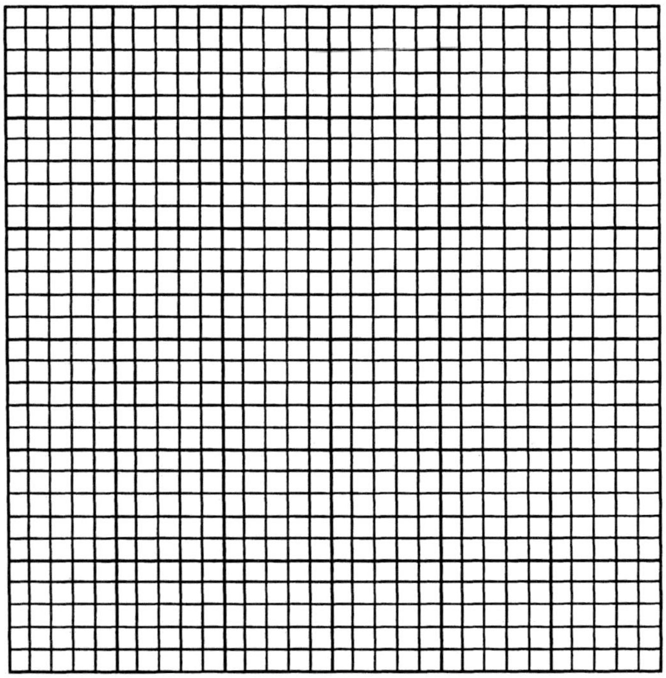

b. _______________

c. How?

d. Why?

Exercise 4. Effect of pH.

a.

Tube	pH	Bubble Height (mm)
1	2	________________
2	7	________________
3	12	________________

Graph: pH (X) vs. Catalase Activity (Y)

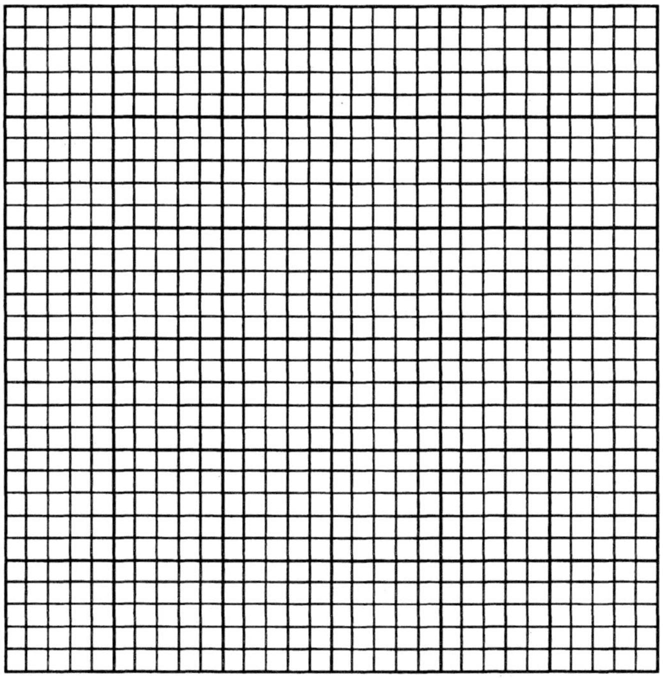

b. Acidic? __________

c. Basic? __________

d. Neutral? __________

e. Optimum? __________

f. Discussion:

Exercise 5. Effect of Temperature.

a.

Tube	Temp. (°C)	Bubble Height (mm)
1	0	____________________
2	25	____________________
3	100	____________________

Graph: Temperature (X) vs. Catalase Activity (Y)

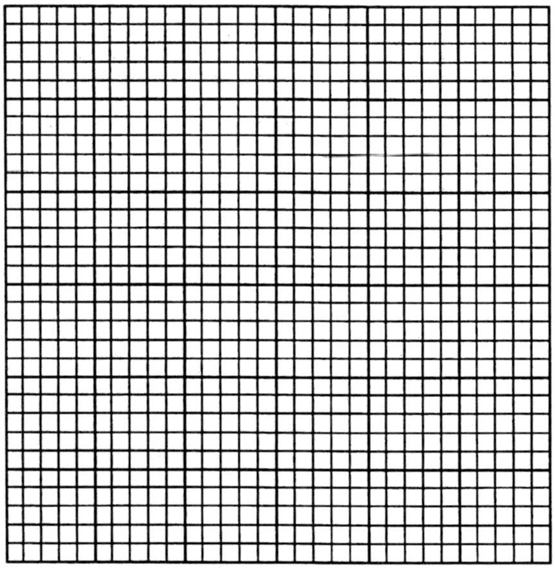

b. Optimum? ________________

c: Discussion:

Exercise 6. Enzyme Inhibition.

a.

Tube	Additive	Bubble Height (mm)
1	water	____________________
2	copper	____________________

b. Effect:

c. Explanation:

Photosynthesis

Introduction

Green plants occupy a unique and important place in the biological world. With few exceptions (certain bacteria and cyanobacteria), green plants are the only living things which are capable of producing foods from non-foods. Such food production is accomplished through the biological process known as "photosynthesis". This term is quite appropriate in that it expresses the concept that plants harvest sunlight energy ("photo") to make food ("synthesis"). It is through photosynthesis that energy first enters the living world. Hence, plants hold their position of importance among living things by virtue of the fact that all other organisms get their energy from the plants (either directly through consumption or indirectly through eating or decomposing other plant consumers). A general chemical equation for photosynthesis could be written like this:

$$6\,CO_2 + 6\,H_2O \xrightarrow{\text{Light}} C_6H_{12}O_6 + 6\,O_2$$

The equation depicts the light-driven consumption of carbon dioxide and water resulting in the production of sugar and the release of oxygen to the atmosphere.

Objectives

Upon completing your study of this unit, you should be able to:

1. Write a general, balanced equation for photosynthesis.
2. List the two major characteristics of light.
3. Specify the range of wavelengths which constitutes visible light.
4. List, in order of increasing wavelength, the colors of light present in photosynthetically active radiation.
5. List, in order of increasing energy content, the colors of light present in photosynthetically active radiation.
6. Describe the role and importance of CO_2 and light in photosynthesis.
7. Define the term "pigment" as it relates to photosynthesis.
8. List three ways in which energy may be released from pigment electrons.
9. Identify reflected colors of light based on the color of a surface.
10. Describe a simple method for the separation of plant pigments.
11. Determine the relative solubility of plant pigments in a solvent by means of paper chromatography.
12. Predict the color of a light beam of a particular wavelength.
13. Graph the relative absorption spectrum of a plant pigment from raw transmission data.

14. List the four photosynthetic pigments observed during this laboratory period.
15. Predict which wavelengths and colors of light are best and which are worst for driving photosynthesis.

The Role of Light

Since light is such an important feature of the photosynthetic process, let us consider some characteristics of light which are pertinent to our study. First we should realize that light has a particulate property. That is to say that light exists in discrete, measurable units called photons and therefore can be quantified. It is through reception of quantities of photons with man-made photocells that we are able to convert sunlight (solar) energy into electricity or other usable forms of energy. The amount of energy produced per unit time is dependent on the "intensity" of the light (number of photons received per unit time). Similar reception of photons by green plants drives the process of photosynthesis.

Second, we must also understand that light travels in waves similar to sound or radio waves. The light which is emitted by the sun and is visible to humans is made up of many waves varying in length (from wave crest to crest) from around 400 to 700 nm (1,000,000 nanometers = 1 millimeter). These same wavelengths also uniquely constitute ***photosynthetically active radiation***, and are the only effective wavelengths for driving photosynthesis.

If we could separate these waves of light from one another, we would find that each wave would exhibit a color determined by its specific wavelength. In other words, white light is actually made up of many colors mixed together.

Note: Exercises 1 and 5 involve considerable waiting periods, so begin them both now. Divide the responsibilities among the members of your group. However, be sure that each person understands both procedures.

Exercise 1. Effects of Light Intensity

Place 50 ml of tap water in a small beaker and add 20 drops of phenol red (a pH indicator) so that it turns reddish-pink. Obtain five 20 ml test tubes and fill one of them half full from the beaker. Place it aside for later comparison. Next, using a soda straw, bubble your breath into the water remaining in the beaker for *only a second or two* until a distinct orange (not peach yellow) color appears. This change in color is due to the mixing of CO_2 from your breath with the H_2O to form carbonic acid (H_2CO_3).

Fill the other four test tubes with equal amounts of the acid solution from the beaker. Mark the tubes 1, 2, 3, and 4 with a grease pencil. Add one pre-cut two-inch-long sprig of *Elodea* to each of tubes 1, 2, and 3 (Do not pull leaves off of sprigs). Seal all five tubes with stoppers. Cover tube 1 completely with aluminum foil and place tubes 1, 2, and 4 approximately 20 cm from a 150W light bulb. Place tube 3 about 50 cm from the same light source and record the time on your answer sheet. Check the tubes which are exposed to the light every 10 minutes and note the color changes which take place over the next 30-60 minutes. As tubes 2 and 3 change to the original pink color (compare to the unlabeled tube) record the time and calculate the time necessary for the change to take place.

Unwrap tube 1 and consider your results. ●a) What caused the color change in tubes 2 and 3? ●b) How does this change indicate that photosynthesis occurred? Hint: consider the equation for photosynthesis at the beginning of this chapter. ●c) What else could we have measured as an indication of photosynthesis? ●d) How do the results from test tube 1 make it clear that light is necessary for photosynthesis? ●e) From a comparison of tubes 2 and 3, what can you conclude about the effect of light intensity on photosynthesis? ●f) Why was no color change observed in tube 4?

Exercise 2. Diffraction of White Light.

Obtain a diffraction grating spectroscope from the instructor. With the slit end directed at a light source, look through the diffraction grating window at the darkened inside wall of the tube. As the white light passes through the narrow slit, it is diffracted (or bent) so that the waves of shorter length are projected closest to the slit end of the scope, while the longer wavelengths are projected closer to the viewing end. ●a) What colors can be observed in the spectroscope? ●b) Which color has the longest wavelength? ●c) The shortest?

Use the spectroscope to view both incandescent and fluorescent light sources. ●d) What differences in color visibility or intensity do you notice between the refraction patterns of these two sources? ●e) What explanation can you offer for these differences?

Exercise 3. Colors and Their Wavelengths

As soon as you have completed Exercise 2, consider the demonstration of color and wavelength which is set up using the spectrophotometer. A piece of beveled chalk should be placed in the bottom of an empty cuvette in the sample chamber. By viewing the chalk while slowly turning the wavelength dial, the wavelengths which correspond to certain colors can be determined. Start with the dial at 700 nm and turn until the sharpest red color is seen. ●a) Record this wavelength on your answer sheet. Continue this procedure for each of the colors listed on the answer sheet.

Plant Pigments

How are plants able to capture and hold sunlight energy? When a beam of sunlight strikes the surface of a leaf, electrons within certain molecules are "excited" (or filled with energy) by particular wavelengths of light. This excitation process is called light or energy absorption. Molecules which are capable of absorbing light energy are known as ***pigments***. Wavelengths of light which are not absorbed by a pigment are reflected and transmitted (allowed to pass through) giving the pigment its apparent color. In other words, green plants are green because they reflect green light while the other wavelengths are absorbed by the various plant pigments.

Light energy absorbed by the pigment molecules is subsequently released after being stored in the electrons for a mere fraction of a second. The energy may be released as heat (resulting in a warming of the leaf) or re-emitted as light. Re-emitted light is called ***fluorescence*** and always contains less energy than the light which was originally absorbed. The absorbed energy may also be used in the formation of chemical bonds between small molecules resulting in the ("photo") synthesis of larger, more complex molecules.

Exercise 4. Wavelength Absorption and Fluorescence.

Obtain a sealed tube or flask of dilute plant extract and move it in front of the slit in the diffraction grating spectroscope while viewing a light source. ●a) What colors are visible? ●b) How does this differ from your observation in Exercise 2? ●c) Offer an explanation for the difference.

Do not use the spectroscope for this next procedure. Shine a light source on a flask of concentrated plant extract while observing the flask from the side. The color you see, other than green, is fluorescence. ●d) What color is the re-emitted light? ●e) From the information provided in the preceding paragraphs, would you say that red light is higher or lower in energy than other visible wavelengths? ●f) Considering your findings in Exercises 3 and 4, is the quantity of energy in light *directly* or *inversely* proportional to wavelength?

Exercise 5. Separation of Photosynthetic Pigments.

Obtain a sheet of chromatography paper on which concentrated plant extract has been applied. Carefully bend the paper into a cylinder so that the green line forms a circle at one end and hold the edges so that they nearly touch while your lab partner staples the edges together at the top and bottom of the cylinder. Bring a glass chromatography jar containing about 1 cm of chromatography fluid to your lab desk.

Place the paper cylinder into the center of the jar so that the green line is down. Be sure the chromatography fluid does not cover the pigment line. Place the lid on the jar and allow the chromatography solvent to advance up the paper cylinder. As this occurs, the pigment molecules within the green extract solution will begin to dissolve into the chromatography solution. The most soluble pigments will move quickly up the paper while the less soluble pigments will lag behind resulting in separation of the major pigments present. Do *not* move or pick up the jar until the process is complete.

While the solvent front is advancing (30 to 40 minutes), resume your reading at the top of the second page of this laboratory. Keep an eye on your chromatograph to be sure the solvent doesn't reach the top of the paper.

RETURN TO THIS POINT

When the solvent front is approximately 1 cm from the top of the cylinder, remove the cylinder from the jar and allow to air dry for a few minutes. Do not empty the jar, but simply return it to the supply counter where you got it. Unstable the paper and examine the chromatograph to determine how effectively the pigments have been separated. You should be able to distinguish five separate bands of pigment (some may be almost too faint to see). The uppermost band is not important for our study and should be ignored. The second band will be a light yellow color and is a ***carotene***, the third (also yellow) ***xanthophyll***, the fourth (blue-green) ***chlorophyll b***, and the bottom band should be ***chlorophyll a*** (brownish-green).

●a) Which pigment is most soluble in the chromatography solution? ●b) How far apart are the top and bottom bands (from the top of one to the top of the next)? ●c) How far apart are the two bottom bands? ●d) Which two pigments are probably most similar structurally? Carefully cut out the pigment bands and place them in the appropriately labeled test tubes at the front of the lab.

Exercise 6. Photosynthetic Pigment Absorption Spectra.

At convenient intervals during the lab period after your chromatography system is set up, your instructor will ask the class to come to the front by groups (one lab group at a time) to determine pigment absorbance with the spectrophotometer. Until your group is called, continue working on the other exercises.

Following the instructions provided on the spectrophotometer, determine the absorbance of a single wavelength (designated by instructor) for each of the four pigment solutions provided. The pigment solutions include carotene, xanthophylls, chlorophyll a, and chlorophyll b. The solutions were prepared by cutting apart the colored bands on chromatography paper produced in the previous exercise.

Record your data on the chalkboard in chart form along with the rest of your class so that the absorbance at each wavelength is available to the entire class. ●a) Fill in the chart on your answer sheet using the class data on the board. ●b) Using the data from the completed chart, graph the absorbance spectra for each pigment on your answer sheet.

Name ______________________

Number ______________________

Photosynthesis

Exercise l. Effects of Light Intensity.

Time at beginning of experiment: _________

Time at end point: Tube 2 _________ Tube 3 _________

Time required for change:Tube 2 _________ Tube 3 _________

a.____

b.____

c.____

d.____

e.____

____

f.____

Exercise 2. Diffraction of White Light.

a.____

b. ______________ c. ________

d.______________________________ e. ______________________________

Exercise 3. Colors and their Wavelengths.

a. Wavelengths: Red ______ nm, Orange ______ nm, Yellow ______ nm,

Green ______ nm, Blue ______ nm, Violet ______ nm

Exercise 4. Wavelength Absorption and Fluorescence.

a.____

b.____

c.____

d. __________ e. _______________

f. ____________________________

Exercise 5. Separation of Photosynthetic Pigments.

a. ____________________________

b. _____ cm. c. _____ cm.

d. __________________________ and ____________________________

Exercise 6. Photosynthetic Pigment Absorption Spectra.

a. Wavelength	400	450	500	550	600	650	700
Carotene	____	____	____	____	____	____	____
Xanthophyll	____	____	____	____	____	____	____
Chlorophyll a	____	____	____	____	____	____	____
Chlorophyll b	____	____	____	____	____	____	____

b. Absorption Spectra

Carotene (o)
Xanthophyll (x)

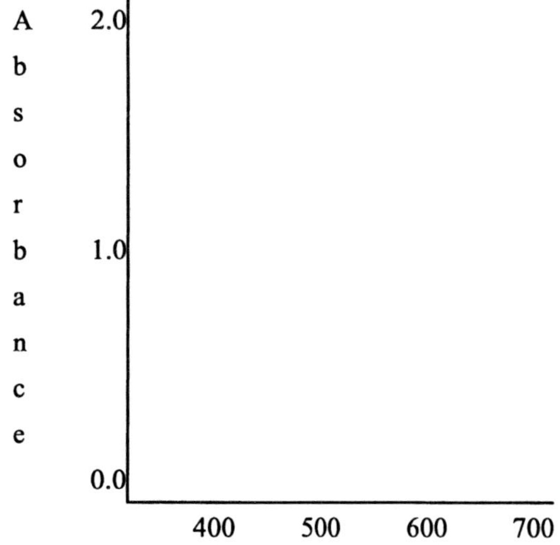

Chlorophyll a (o)
Chlorophyll b (x)

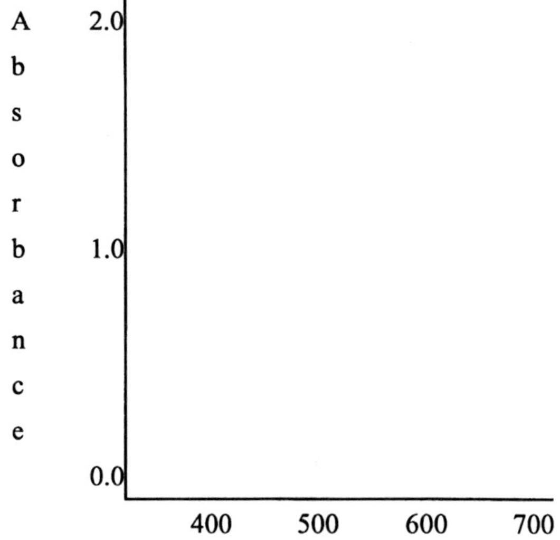

Fermentation and Respiration

Introduction

Living organisms require a source of energy to perform the functions necessary for life. On earth we are dependent upon the sun for our energy. The energy of sunlight is absorbed during photosynthesis by plants, algae, cyanobacteria, diatoms, etc., and is stored in the chemical bonds of sugars which are the primary products of photosynthesis.

Non-photosynthetic organisms such as animals are dependent upon this "packaged sunlight" since we cannot tap the sun's energy directly. Therefore we eat food and break down the chemical bonds to release the energy. This breakdown is accomplished by enzymes in the cell, and the released energy is transferred to the intracellular energy molecule ATP. ATP then serves as an immediate source of energy for most cellular functions such as cell division, protein synthesis, movement, etc.

There are two ways that cells break down energy sources (food). These are fermentation and respiration. Fermentation involves a partial breakdown of an energy source resulting in a partial release of energy and the formation of a variety of different organic molecules as end products. Respiration involves a complete breakdown of an energy source resulting in maximum release of energy and the formation of carbon dioxide. Respiration is normally ***aerobic*** (requires oxygen), while fermentation is typically ***anaerobic*** (does not involve oxygen consumption).

So in the absence of oxygen fermentation continues, but respiration stops. Small unicellular creatures that can perform both fermentation and respiration would simply switch to fermentation and survive nicely. However, animals cannot survive without respiration. Such a deprivation of oxygen would result in cell damage and eventually death. You can get a good idea of how long you can survive without oxygen by holding your breath. In this laboratory we shall examine these two very important energy-yielding metabolic options: fermentation and respiration.

Objectives

Upon completing your study of this unit, you should be able to:
1. Write the reaction for alcoholic fermentation.
2. Write the reaction for aerobic respiration.
3. Describe an experiment to test the relative fermentation rates of various sugars.
4. Identify the product of fermentation most important in the brewing industry.
5. Identify the product of fermentation most important in the baking industry.
6. Identify the control in a given experiment and state its purpose.

7. List the advantages to an organism of respiration over fermentation.
8. Explain the purpose of potassium hydroxide in the respiration experiment.

Fermentation

There are several common forms of fermentation, but certainly one of the most significant is alcohol fermentation. This fermentation is carried out by a variety of microorganisms among which the unicellular fungus, yeast, is probably the most important. Yeast is used by humans to convert fruit juices into alcoholic beverages. Sometimes the ethyl alcohol (ethanol) is separated and used for industrial purposes or as a gasoline additive, forming "gasohol". What many people do not realize is that the same yeast is used by the baking industry to produce carbon dioxide gas which makes dough rise. In ethanol fermentation, glucose sugar ($C_6H_{12}O_6$) is converted into carbon dioxide (CO_2) and ethanol (C_2H_5OH). During the process, enough energy is released to form two molecules of ATP for each molecule of glucose fermented. The following equation summarizes alcohol fermentation:

$$C_6H_{12}O_6 + 2ADP + 2P \rightarrow 2CO_2 + 2\,C_2H_5OH + 2ATP$$

Exercise 1. Alcohol Fermentation.

Note: Because of the time involved, half of your group should set up Exercise 1, and half should set up Exercise 2.

In this exercise we'll use the common baking yeast, *Saccharomyces cerevisiae* to study fermentation. We can measure the rate of fermentation by the rate at which gas is produced in a closed container. ●a) What gas is produced during ethanol fermentation? ●b) How many molecules of ATP are produced from each molecule of glucose fermented?

We will examine the relative fermentation rates for solutions of glucose (a simple sugar), sucrose (a disaccharide made of two simple sugars, glucose and fructose), and corn syrup (a diverse mixture of large complex sugars and smaller simple sugars). ●c) Realizing that glucose is the starting substrate for the ethanol fermentation and that complex sugars must be broken down and converted to glucose before they can be fermented, propose a hypothesis (prediction or explanation) as to which solution will be fermented most quickly and which most slowly. After recording your hypothesis, get four clean beakers. Use these to prepare the yeast/substrate mixtures by adding the following to separate beakers. In order to distribute the yeast cells evenly in the suspension, swirl the flask before pouring.
#1: 10 ml yeast suspension + 10 ml glucose,
#2: 10 ml yeast suspension + 10 ml sucrose,
#3: 10 ml yeast suspension + 10 ml corn syrup, and
#4: 10 ml yeast suspension + 10 ml water (This serves as a control).

Mix each beaker well.

Label four fermentation tubes #1 – #4 and fill each with the contents of the corresponding beaker. Tilt the fermentation tubes so that the tubular portion completely fills with liquid— excluding all air from the tube. Return the tubes to a standing position, and place in an incubator or water bath at 37°C. Allow them to remain undisturbed until the liquid level in any one of the tubes descends at least 50 mm.

Note: While you are waiting, familiarize yourself with the procedure for Exercise 2. All students are responsible for understanding both exercises.

Using a metric ruler, measure the final amount of gas in each tube (in mm), and ●d) record the data on the answer sheet. ●e) Do your data support your hypothesis? ●f) Explain. ●g) If not, write a modified hypothesis that is consistent with your data. ●h) What is the purpose of tube #4?

Respiration

The following equation summarizes the respiratory breakdown of glucose into carbon dioxide and water:

$$C_6H_{12}O_6 + 6O_2 + 38ADP + 38P \rightarrow 6CO_2 + 6H_2O + 38ATP$$

Respiration occurs in the mitochondria while fermentation occurs in the cytoplasm of a cell. As discussed earlier, respiration is an oxidation process (electrons removed from glucose). The electrons are passed through the electron transport system finally reducing oxygen to form water. Respiration did not occur in the first experiment because the oxygen in the tubes was quickly used up, and there was no other final electron acceptor for the electron transport system in the then anaerobic tube.

During respiration oxygen (O_2) is consumed, and carbon dioxide (CO_2) is produced by cells. While many kinds of cells could be used, we'll use germinating pea seeds which are readily available, inexpensive, and have a relatively high rate of respiration. In this study, we will determine respiration rates by measuring oxygen consumption. To do this, carbon dioxide must be trapped and removed from the air so we can detect oxygen utilization by a decrease in total gas volume. We will use potassium hydroxide (KOH), a chemical which absorbs carbon dioxide gas, to remove the CO_2 from the gas phase. The only precaution here is that the KOH must be physically separated from the cells being studied, because it is harmful to living cells... pea cells or *your* cells!

The principle we are using is similar to the use of a soda straw. When you "suck" on a straw, what you are actually doing is removing gas from the straw. Liquid then moves up into the straw due to the difference in air pressure inside the straw relative to that outside. We will construct a simple respirometer (meter for measuring respiration) that is based on this same principle.

Exercise 2. Pea Seed Respiration.

Fill two 125 ml flasks to about 75 ml with the germinating pea seeds — these seeds are comprised of living, respiring plant cells. These will be your experimental flasks. Fill a third flask to about 75 ml with ungerminated seeds or plastic beads. This flask will be your control which contains no respiring cells. You will always need to correct your experimental data by subtracting any changes in gas volume after experimental measurements are complete.

Place a *closed* vial of KOH pellets in each flask. **Caution:** Do not touch the KOH. It is extremely caustic and will injure your skin. If you do touch some KOH, rinse your skin thoroughly under running water.

Get three rubber stoppers, each of which has a bent graduated pipette (respirometer tube) passing through it. Each stopper will also have a second short glass tube (vent) with a short piece of soft tubing attached. With a bit of water on your finger, moisten the edges of the stoppers and insert them tightly into the three flasks.

Leave one experimental flask and the control on your lab table at room temperature. Place the second experimental flask in an ice bath at °C, and do not remove it until the end of the experiment.

●a) Propose a hypothesis predicting what you think will be the effect of temperature on respiration rate.

Note: Wait *fifteen minutes* for the temperatures inside and outside of the flasks to become equal. While you are waiting, familiarize yourself with the procedure for Exercise 1. All students are responsible for understanding both exercises.

After fifteen minutes have elapsed, using an dropper or dropper bottle, add a drop of dye to the end of each respirometer tube. About 1 cm of dye in the respirometer tube will be sufficient. Then close the clamp on the tubing attached to the vent on the rubber stopper of each flask. Do not handle the flask again until the end of the experiment.

Using the 0.1 ml graduations on the respirometer tube, measure the volume of O_2 consumed in milliliters by tracking the movement of the dye droplet from the end of the respirometer tube. Measurements should be made at regular 1 minute intervals. If the dye is barely moving, increase the intervals to 2 minutes or longer. You should try to make at least 5 measurements over a time period of 15 minutes.

●b) Each time you measure, record on the data sheet the volume consumed for all three tubes and the time elapsed. It would be best if one person followed the time, and at his or her signal, three separate individuals measured the three tubes simultaneously.

The volume consumed (if any) by the control will have to be subtracted from the experimental tubes each time a measurement is taken. ●c) Why? ●d) After the experiment is completed, plot the volume consumed (as a measure of respiration) versus time in minutes. ●e) Calculate the respiration rate for each of the experimental tubes by dividing the total volume (in milliliters) by the total time (in minutes). This results in the average respiration rate in milliliters (ml) per minute (min). ●f) Offer an explanation for your results. (Hint: refer to Exercise 5 in the "Enzymes" laboratory.) ●g) Did the data support your hypothesis? ●h) If not, propose a modified hypothesis consistent with the data.

Exercise 3. Respiration vs. Fermentation.

●a) How many ATP's are produced from the respiration of one molecule of glucose (Consult equation for respiration)? ●b) How does that compare with the ATP's produced from the fermentation of glucose? (consult equation for fermentation) ●c) Why would it be advantageous to an organism to carry out respiration if oxygen or some other final electron acceptor is available?

Return the germinated pea seeds and beads to the appropriate containers.

Name ______________________________

Number ___________________________

Fermentation and Respiration

Exercise 1. Alcohol Fermentation.

a. ________________ gas

b. ________________ ATPs

c. Hypothesis:

	Tube #1	Tube #2	Tube #3	Tube #4
d. Final gas (mm)	________	________	________	________

e. ________________

f.____

____

____

____

g.____

____

____

h.____

____

Exercise 2. Pea Seed Respiration.

a. Hypothesis:

b.

Time	Control	Flask #1	Flask #2	Corrected* Flask #1	Corrected* Flask #2

*Corrected flask = uncorrected flask control flask

c. Why?

d. Graph of results:

	Flask #1	Flask #2
e. Total Corrected Volume (ml)	_______	_______
Total Time (minutes)	_______	_______
Corrected Respiration Rate (ml/min)	_______	_______

f. Explanation:

g. Did the data support your hypothesis (yes or no)?

h. Modified hypothesis (if necessary):

Exercise 3. Respiration vs. Fermentation.

a. ________ ATPs

b. Comparison:

c. Advantages:

Kingdoms Prokaryotae and Protista

Introduction

Aristotle and his contemporaries classified all living organisms as either plants or animals. This two-kingdom system was adequate for more than two millennia; however, during the 19th century it became evident that there were clearly some creatures that were neither plants nor animals. Microscopic unicellular organisms seemed particularly out of place in both kingdoms. The German biologist Ernst Haeckel (1834-1919) therefore proposed a third kingdom which he called the ***Protista*** to accommodate all microorganisms that seemed to be distinct in character from both plants and animals.

As microscopes became more sophisticated in the 20th century, evidence for a fundamental distinction between two broad classes of cells began to accumulate. One group of microorganisms was characterized by its relative simplicity of internal structure. These single-celled creatures had no true nucleus separated from the rest of the cell by a nuclear membrane. These cells were also without ***organelles***: membrane-bound structures within the cell which independently perform specific functions. For example, they had no chloroplasts, mitochondria, centrioles, lysosomes, or most of the other organelles typical of other living cells.

To Eduard Chatton, these distinctions in cell size and complexity represented a major discontinuity in the biological world. In 1937 he proposed separate terms for these two kinds of cells. The smaller, simpler type of cells were termed ***prokaryotic*** (Greek: *pro* = before + *karyon* = nut or kernel, i.e., nucleus; lit. cells without nuclei). They include those cells that are commonly referred to as **bacteria**. The more complex cell type, characteristic of all other organisms whether uni- or multicellular was referred to as ***eukaryotic***, i.e. "true nucleus".

During this same period of time, botanists came increasingly to view the molds and fungi as being metabolically and structurally quite distinct from members of the Plant Kingdom.

In order to accord appropriate significance both to prokaryotes and to the Fungi as separate groups of organisms, an American biologist, R.H. Whittaker, proposed (1959) a five-kingdom classification system. This system included the three existing kingdoms (Plant, Animal, and Protista) plus two new ones. (1) The Fungi (molds, yeasts, and mushrooms), all of which have eukaryotic, non-photosynthetic cells, were removed from Kingdom Plantae and given their own designation: ***Kingdom Fungi***. (2) the Prokaryotes which include all "simple-celled" organisms without a nucleus or organelles were removed from Kingdom Protista and given their own Kingdom designation. In 1974, the editorial board of *Bergey's Manual of Determinative Bacteriology*, gave this last

Kingdom the name ***Kingdom Prokaryotae.***

Once the prokaryotes were removed from Ernst Haeckle's Kingdom Protista, the groups of organisms that remained were highly diverse—they defy a simple unifying classification scheme. They include unicellular and colonial-unicellular, photosynthetic (autotrophic) and non-photosynthetic (heterotrophic) organisms. They are mostly aquatic (live in water) and microscopic. It is these creatures that so often captivate the student during a microscopic examination of a drop of pond water. Kingdom Protista functions somewhat taxonomically as a catch-all kingdom for organisms that don't fit anywhere else. The plant and animal kingdoms include multicellular photosynthetic and multicellular non-photosynthetic eukaryotic organisms respectively. The Kingdom Prokaryotae includes all prokaryotic creatures. The Kingdom Fungi... well, they're just fungi with all their uniqueness. Everything else belongs to the Kingdom Protista! While the protists are relatively simple compared to plants and animals, you will see that even a single-celled organism can be quite complex.

Kingdom Prokaryotae

Phylum Archaea – methane-producers, salt-loving, thermophilic forms, diverse habitats
Phylum Bacteria - common soil, water, forms; plant, animal pathogens

Kingdom Protista

Phylum Mastigophora — heterotrophic, move by flagella, e.g., *Trypanosoma*
Phylum Ciliophora — heterotrophic, move by cilia, e.g., *Paramecium*
Phylum Sarcodina — heterotrophic, move by pseudopodia, e.g., *Amoeba*
Phylum Chrysophyta — autotrophic, diatoms
Phylum Euglenophyta — heterotrophic and autotrophic, photosynthetic euglenoids
Phylum Chlorophyta — autotrophic, green algae; photosynthetic unicellular and multicellular forms which many consider to be plants.

In this lab we begin our "five-kingdom study" of diversity with examples of single-celled and colonial forms taken from Kingdoms Prokaryotae and Protista.

Objectives

Upon completing your study of this unit, you should be able to:

1. List three differences between prokaryotic and eukaryotic cells.
2. Name and draw three basic prokaryotic cell shapes.
3. Name the organisms that cause strep throat and tetanus.
4. List four general ways diseases can spread.
5. Identify from a specimen or drawing an amoeba and pseudopodia.
6. Identify from a specimen or drawing a paramecium; describe eating and digestion process.

7. Describe the function of a contractile vacuole.
8. Identify the method of locomotion for each of the first three phyla of the Protista.

Prokaryote versus Eukaryote

As discussed in the Introduction, there are many significant differences between prokaryotic and eukaryotic cells. One can often determine whether a cell is prokaryotic or eukaryotic by microscopic examination for size, presence or absence of a nucleus, and types of organelles. In the first exercise a photosynthetic eukaryote will be compared with a photosynthetic prokaryote.

Exercise 1. Comparison of Prokaryotic and Eukaryotic Cells.

Obtain prepared slides or make wet mounts of both *Spirogyra* and *Oscillatoria* which are filamentous, photosynthetic algae. Compare the two cell types for size, presence or absence of a nucleus, and presence or absence of chloroplasts. Examine both organisms with your microscope using the high power objective. ●a) Make a simple sketch of each organism. ●b) Indicate the magnification used for each sketch. Determine which organism is larger and which is smaller and ●c) record "larger" or "smaller" in the space provided for describing size on your answer sheet. Although size is generally an unreliable criterion alone, in conjunction with other observations it can be useful. Prokaryotes, because of their relative simplicity, tend to be smaller than eukaryotes.

Chloroplasts are complex green organelles which perform photosynthesis in eukaryotes. Since the pigments of photosynthetic eukaryotes are concentrated in the chloroplasts, the rest of the cell will appear relatively colorless. In prokaryotes the pigment or color is more evenly distributed throughout the cytoplasm since chloroplasts are absent. Sometimes chloroplasts have distinctive shapes that are useful in identifying eukaryotic algae. Determine which organism has a chloroplast(s) and which one does not. ●d) Record your observations on the answer sheet by writing "present" or "absent" in the space provided for the chloroplasts.

In a previous lab you observed the nucleus of a human cell as a roundish structure near the center of the cell. Examine each of the two algae. Look for a nucleus. Nuclei are hard to find in algal cells. Use the fine focus adjustment to focus up and down through the cell. Be sure you don't miss the nucleus due to its being out of focus. ●e) Record your observations on the answer sheet by indicating "present" or "absent" in the space provided for the nucleus.

Draw each alga in the space provided on the answer sheet and label the chloroplast and nucleus (if visible). ●f) Write your conclusion about each alga as to whether it is a prokaryote or a eukaryote.

Kingdom Prokaryotae

Bacteria

Bacteria are found everywhere... soil, water, air, and in and on plants and animals. Most are harmless and in fact are important organisms in decomposition and nutrient cycling. Some bacteria are able to survive on plants and animals and can cause abnormal problems that result in disease.

Although single cells are too small to be seen with the unaided eye, the colonies that result from a single cell or a few cells repeatedly reproducing by fission (division) on a solid growth medium can be readily examined. In fact, colony characteristics can be very useful in identifying bacteria. The following is a list of a few useful colony characteristics:

Criterion	Possibilities
shape	round: 
	irregular: 
texture	dull, shiny or mucoid
color	many different colors are found
edge	entire (smooth): 
	undulate (wavy): 
	lobate (lobed): 

Bacteria are often grown on a nutrient growth medium that is solidified to a firm jellylike consistency with a gelling agent called agar. The nutrient agar medium is mixed, sterilized by heating, and poured into a suitable container where the medium solidifies as it cools. The container most often used is a round covered dish called a petri dish (or plate) named for the bacteriologist who first developed its use in the late 1800's.

Prokaryote Cell Shape

Prokaryotic cells exist for the most part in one of three basic shapes: coccus (pl. cocci), bacillus (pl. bacilli), or spirillum (pl. spirilla). Cocci are spherical and may remain attached together after fission to form pairs (diplococci), clumps or clusters (staphylococci), or chains (streptococci). A particular cellular arrangement is characteristic of a species and is often useful in bacterial identification. Bacilli are cylindrical cells and usually occur singly, in pairs, or in chains. Spirilla are curved cells that range from comma-shaped, to "S"-shaped, to truly spiral shapes. While spirilla usually occur singly, they may sometimes form short chains.

While most bacteria are not harmful to man and in fact many are beneficial, some can grow in the human body and cause disease. Disease-causing bacteria are called pathogens. A few of these will be used to examine the three basic prokaryotic cell shapes:

Organism	Basic Shape	Human Disease
Streptococcus pyogenes	coccus	Streptococcal sore throat, scarlet fever, impetigo
Clostridium tetani	bacillus	tetanus (lockjaw)
Spirillum volutans	spirillum	non-pathogenic (lives in pond water)

Exercise 2.

Examine prepared slides of each of the three basic prokaryotic cell shapes under high power and record your results on the answer sheet. ●a) Draw each organism, and ●b) identify it as to which of the three basic shapes it represents, i.e., coccus, bacillus, or spirillum.

Infectious Pathogens

Pathogens, disease-causing organisms, can be spread among humans by direct contact (kissing, sexual interaction, touch, etc.), airborne transmission (coughing, sneezing), food, or water. Since bacteria are too small to see with the unaided eye, they are present all around you without your being aware of it. In fact we have relatively harmless bacteria that live in us (intestine, mouth) and on us (skin) all the time. In the next exercise you will have an opportunity to grow some of these bacteria.

Exercise 3.

Obtain a petri dish containing fresh sterile growth medium. Remove the lid and inoculate the plate possibly in one or several of the following ways:

a. Kiss the agar.
b. Touch with fingers.
c. Leave open to air in room for 5-10 minutes.

If none of these methods appeals to you, be creative and try something else. You may even want to try an experiment such as comparing touch inoculations with washed and unwashed hands or cough inoculations with and without a handkerchief over the mouth. Use half the plate for each inoculation. If you inoculate in more than one way, cover the part of the plate not being inoculated at a given time with a piece of paper to keep "germs" out. With multiple inoculations it's a good idea to sector or divide your plate into even areas by marking the bottom with a wax pencil or marking pen. Label each area to identify the method of inoculation.

After inoculating the plate, replace the lid, tape the lid shut, label the plate on the bottom with your name or initials and lab section, and place the plate upside down in the location designated by your instructor. At the beginning of the next lab period you will be given time to examine the plate for bacterial growth. Do *not* open the plates. Do *not* remove the plates from the lab.

After several days, observe your plate and ●a) sketch a drawing of the plate with its developing colonies. For the larger colonies on the plate, write a 3 – 4 word description using terms for colony morphology given above. Also indicate ●b) the method of inoculation, ●c) days of incubation, and ●d) any conclusions you can draw from your experiment. You may find it interesting to examine the plates of others in the class who used a different method of inoculation. ●e) Why do you suppose you were told not to open the plates or to remove them from the lab?

Kingdom Protista

The Animal-like Protists

Members of this Kingdom have been generally grouped into the animal-like protists (sometimes called protozoa), plant-like protists (algae) and the fungal-like protists (certain molds).

The animal-like protists comprise an extremely large and diverse group including three phyla. There are probably about 65,000 known species of the animal-like protistans. They are single cells, but what marvelous cells they are! In that single cell, all of life's processes must occur. Each for the most part, except in a few colonial forms, is on its own and must perform all functions necessary for life. It has been suggested that the simplest protozoan cell is much more complex than the most complex human cell, largely because the cells of multicellular organisms (plants and animals) are specialized to perform only one or a few of the functions important to the whole organism. We shall examine members of each class of the Phylum Protozoa, which traditionally have been classified according to their means of locomotion.

Phylum Sarcodina

The most familiar organism in this group is the amoeba (Greek *amoibe* = change). Although a variety of shelled organisms are common in this class, amoebae are irregular in shape and are not bound by a cell wall. Two internal structures are usually visible in the amoeba: ***food vacuoles*** (darker spherical sacs containing partially digested food) and ***contractile vacuoles*** (roughly spherical organelles that enlarge and contract to expel water from the cell). Amoebae move by ***pseudopodia*** (Greek *pseudo* = false + *podia* = feet). A pseudopod is an evagination or "pushing out" of the cell in a given direction. Essentially, the cytoplasm and membrane form a lobe-like protrusion from the cell which gradually enlarges while the membrane and cytoplasm simultaneously pull in from the rear. This pushing out in the front and pulling in from the rear results in a net movement in the forward direction.

Amoeboid motion is accomplished by sol(fluid)/gel(solid) reversals in the cytoplasm. Gelation in the posterior pushes liquid cytoplasm forward extending the pseudopods. Once extended, the cytoplasm along the sides (ectoplasm) gels while the cytoplasm toward the center (endoplasm) remains liquid and pushes the pseudopod forward. Contraction of the posterior is then accomplished by reversion to the liquid state. By extending pseudopods around prey, amoebae engulf food by the process known as ***phagocytosis***. When the food is surrounded by pseudopods and engulfed by phagocytosis it ends up inside the cell surrounded by a membrane sac, the food vacuole, in which digestion occurs. All of this is regulated within the single cell of this "simple" creature.

Exercise 4.

Prepare a wet mount from a culture of amoebae. *Do not move or stir up the cultures!* You should be able to see the amoebae as tiny white spots on the bottom of the culture container. Using a stereo microscope or the naked eye, you should be able to draw an amoeba into a dropper and deposit it on a slide along with a minimal amount of water. The trick is to only *slightly* depress the dropper bulb *before* inserting it into the water. Then you can release it completely to capture the amoeba in the right amount of water.

Observe the amoeba under *low power* first without adding a cover glass. Observe the pseudopods and the cytoplasm streaming out into them. Look also for food vacuoles and contractile vacuoles. A food vacuole is recognized as a sac which may contain a live creature if recently engulfed. The contractile vacuoles are distinguished by their periodic contractions. These expel excess water from the cell, and when they contract they diminish in size so much that they nearly disappear. After contraction they begin to gradually increase in size as they take in water. You'll need to watch an amoeba constantly for a few minutes to see if any vacuoles are contracting.

Amoebae are nearly colorless so you may have to reduce the light with the diaphragm of your microscope. Sometimes other protozoans may be seen swimming around, and you may be lucky enough to see your amoeba engulf one! Perhaps a drop of *Paramecium* culture will whet his appetite? After seeing what you can without a cover glass, try adding one to the slide, and continue your observations. Amoebae are relatively large three-dimensional creatures, and sometimes they are injured or killed by being squashed by a cover glass. On the other hand, the various vacuoles and other internal structures are better seen with a cover glass in place. ●a) Draw an amoeba on your answer sheet, and label pseudopod, food vacuole, and contractile vacuole.

Phylum Ciliophora

The ciliates come in all shapes and sizes, yet share common means of motility, i.e., cilia. Cilia are relatively short bristle-like appendages that beat in a coordinated rhythm to produce directional movement… no small trick in a single-celled organism like *Paramecium*. The paramecium is a slipper-shaped creature which is completely covered with cilia and has a groove, the oral groove, on one side. The oral groove channels food into the cytostome or mouth. The cytostome opens into the cytopharynx which is actually an open sac where food accumulates. The oral groove is located centrally in the cell, and the cytostome and cytopharynx are seen as a sac or clear area at one end of the oral groove. Food vacuoles form from the cytopharynx by a membrane-pinching process similar to the phagocytosis of the amoeba. You may observe one or more food vacuoles with more-or-less-digested food in them. Paramecia will also have one or more contractile vacuoles which periodically contract, excreting water from the cell.

Exercise 5.

Prepare a wet mount of *Paramecium* culture. A drop of methyl cellulose (Protoslo®) mixed in well will slow the organisms down so that you can observe them more easily. You should be able to see the cilia beating. After you have examined your organisms, remove the cover slip and mix in some Congo red-stained yeast cells. Watch the *Paramecium* eat and digest the yeast. The stained cells will pass down the oral groove, through the cytostome and cytopharynx, and into the cell. As they enter the cell they are surrounded by a membrane, forming a food vacuole in which they are digested by enzymes. Congo red is a pH (acid/base) indicator that is blue under acidic conditions and red/orange under basic or alkaline conditions. Determine if any color changes, which would be indicative of pH changes, occur during digestion of the yeast. ●a) Draw *Paramecium* on your answer sheet and label oral groove, food vacuole and contractile vacuole. ●b) Describe any changes in pH as determined by change in color during digestion that you observe.

APPENDIX

The following commentary contains no exercises. It provides information on additional major groups of Protists which may be examined if time allows.

Phylum Mastigophora

The flagellates move by means of long whip-like appendages called ***flagella***. One important group of flagellates are the trypanosomes which are the cause of sleeping sickness and Chagas disease. In some parts of Africa cattle cannot be kept due to the dreaded Nagana caused by trypanosomes. Tsetse flies transmit the trypanosome and spread these diseases by biting first an infected animal and then a healthy animal.

The Algae

The algae comprise all of the protists that are more "plant-like" in form and function. These phyla are photosynthetic—they contain chloroplasts. Some are unicellular. Others are colonial—that is—they are cells that are independent in their most critical functions but that divide and remain physically associated in long filaments or clusters.

Phylum Euglenophyta

Euglena is such a different organism that it is classified in a separate phylum. Euglenae live in slow-moving freshwater streams or ponds where they are so abundant that they give the water a distinct green color. The cell is elongated with a flagellum protruding from an opening, the cytostome, at the tip of one end. As in *Paramecium*, the cytostome opens into a cytopharynx or gullet which widens a bit at the bottom into the reservoir. The flagellum is anchored at the base of the sac-like reservoir and protrudes out of the cell through the cytostome.

Along one side of the reservoir *Euglena* has a light-sensitive eye spot containing reddish-orange pigment which helps it navigate using light as a directional signal. *Euglena* also has a centrally-located nucleus and chloroplasts throughout the cell. Photosynthesis is the primary source of energy, but the euglena can also supplement its photosynthesis by heterotrophic nutrition.

Phylum Chrysophyta

The Chrysophyta or "golden plants" all contain large amounts of a yellow pigment which is important in their photosynthesis. The phylum consists completely of highly geometric "glass"-covered one-celled organisms called diatoms which are very common in both fresh and marine surface waters. Because they are so plentiful in the ocean, and in light of the fact that ocean covers 3/4 of the earth's surface, diatoms are one of the primary photosynthetic creatures on our planet. Diatomaceous earth is formed by the accumulation of the silica shells of diatoms in ancient seas. It is used as an abrasive and in water filtration.

Phylum Chlorophyta

There are over 7000 species of green algae. Most of them are unicellular, colonial, or filamentous forms which live in freshwater. Because there are multicellular forms, the green algae are considered by some to belong to the plant kingdom. Other authorities, however, place them in the protista because there are so many unicellular species. Neither of these classification systems is without merit, but we will deal with them here for convenience.

Name ______________________

Number ______________________

Kingdoms Prokaryotae and Protista

Exercise l.

a. Drawings:	*Oscillatoria*	*Spirogyra*
b. Magnifications	____________X	____________X
c. Size?	____________	____________
d. Chloroplasts?	____________	____________
e. Nucleus?	____________	____________
f. Prokaryotic or Eukaryotic?	____________	____________

Exercise 2.

a. Drawings:	*Streptococcus*	*Clostridium*	*Spirillum*
b. Shape?	____________	____________	____________

Exercise 3.

a. Drawing:

Descriptions:

Colony #1

Colony #2

Colony# 3

b. Method of inoculation:

____

c. Days of incubation: __________

d. Conclusions:

e. Why?

Exercise 4.

a. Drawing of *Amoeba*:

Exercise 5.

a. Drawing of *Paramecium*:

b. pH description:

Kingdoms Fungi and Plantae

Introduction

When Carolus Linneaus first proposed his classification system for living organisms, there were only two major groups or kingdoms proposed – Plantae and Animalia. The development of further technology, especially microscopy, allowed the detection or differentiation of additional major groups. The detection of unicellular organisms by microscopy allowed biologists to describe the prokaryotic bacteria and blue-green algae in the Kingdom Prokaryotae, and the diverse eukaryotic unicellular organisms in the Kingdom Protista. The development of biochemical techniques allowed biologists to recognize that the two types of organisms classified as plants, the fungi and green plants, were as different from one another as either were to animals. Thus, they are recognized today as the Kingdom Fungi (Latin *fungus* = mushroom) and the Kingdom Plantae. In today's lab, you will examine specimens of fungi and the green plants to see some of these differences.

The Kingdom Fungi consist of a diverse group of organisms including familiar multicellular forms like mushrooms, puffballs and bread mold, as well as a few unicellular forms such as yeast. Most fungi have a multicellular body (mycelium) consisting of branching filaments (called hyphae). Most of the fungal mycelium is usually not visible, with many hyphae growing down into the soil or into a food source. Hyphae growing into a food source will secrete digestive enzymes into the food, breaking it down into smaller and simpler organic molecules which can then be absorbed and used for growth and energy. Fungi, like mushrooms, which grow on dead organic matter are termed saprophytes. Others, like the one causing athlete's foot, grows on (or into) living tissue and is termed a parasite. Fungi are important for both of these modes of living. There are many diseases, in humans, animals and plants which are caused by fungi. They are also important decomposers, especially of woody trees, recycling important nutrients back into food chains after organisms die.

The green plants make up the Kingdom Plantae. They have very different structures and life cycles, which is why they are placed in a different Kingdom. While the fungi are heterotrophs in that they derive their food by consuming other organisms, plants are photosynthetic in that they synthesize their food using the energy in sunlight. Plants are sometimes called the green plants because of the universal presence of chloroplasts, the organelles responsible for photosynthesis which contain the green pigment chlorophyll. Fungi and plants are very different in this most basic of biological processes, how they obtain energy for growth and development. Even in areas where they appear similar, fungi and plants exhibit many more fundamental differences. Both fungi and plants build cell walls. However, the structure and composition of these walls is very different. Fungi build cell walls of chitin while plants build them of cellulose.

Objectives

Upon completing your study of this unit, you should be able to:

1. Diagram the typical life cycle of a fungus.
2. For bread mold, be able to identify a zygospore and sporangium, knowing what type of reproduction each represents.
3. For mushrooms, be able to label the stalk, cap and gills.
4. For mushrooms, know what is produced on the gills, and what type of reproduction this represents.
5. Know what fungal buds are, and what type of reproduction they represent.
6. Understand the typical life cycle of plants, involving the alternation of generations.
7. Know whether each life stage in plants represents sexual or asexual reproduction.
8. Know whether each life stage in plants represents the gametophyte or sporophyte generation.

Kingdom Fungi

Let us examine some fungi and see how they complete their life cycles. The body of all fungi is made up of numerous hyphae, those thread-like filaments. The cells of these hyphae are haploid, they contain only one set of chromosomes per cell. The life cycle of many fungi is exactly opposite of that of animals and humans. Human cells are diploid for the entire life cycle except one cell, the gametes (egg and sperm) which are haploid. Fertilization between two haploid gametes reestablishes the diploid chromosome number. In the fungi, the entire life cycle consists of haploid cells, except for one cell. Two haploid gametes, produced by mitosis, unite in fertilization to produce a diploid zygote. In most other kingdoms, a diploid zygote typically grows by mitosis into a multicellular diploid organism. This does not occur in fungi. In most fungi, the zygote immediately undergoes meiosis to produce four haploid meiospores. Each of these haploid spores has the potential to germinate and grow into a new haploid multicellular organism (fungus).

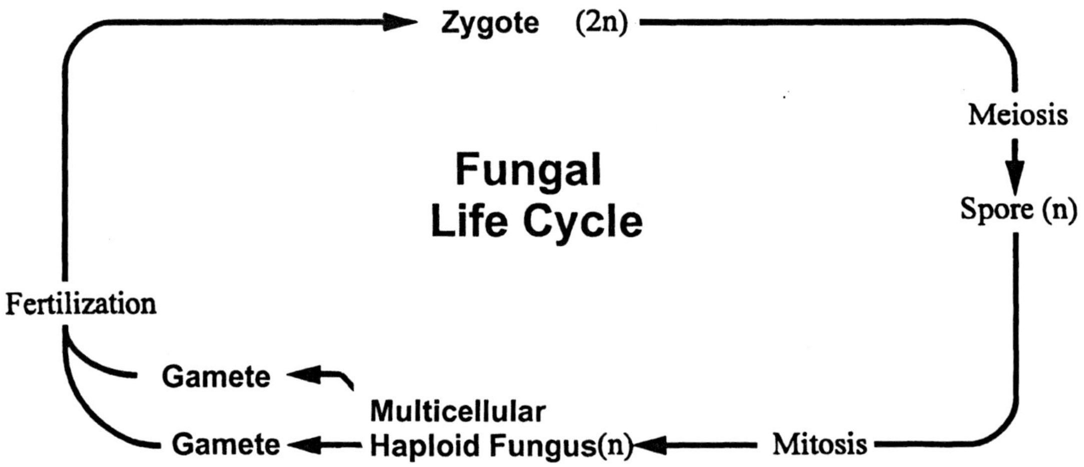

Exercise 1.

Sexual reproduction can clearly be seen in *Rhizopus nigrificans,* the common black bread mold. Obtain a slide marked *Rhizopus* conjugation and examine it under the medium powered objective of your light microscope. The large dark-staining rough-textured spheres are termed zygospores. They represent the zygote, the only diploid cell in the entire life cycle of *Rhizopus*. Zygospores form when hyphae from two different fungi grow close to one another. Hyphae from both fungi develop gametes which fuse to form the zygospore. ●a) On your answer sheet label the zygospore and a gamete from one of the "parent" fungi. ●b) Does the zygospore represent sexual or asexual reproduction? ●c) Is the zygospore diploid or haploid? ●d) Are the gametes diploid or haploid?

Exercise 2

The most familiar fungi are the mushrooms. The hyphae of most mushrooms grow underground. Sexual reproduction occurs when these underground hyphae from two mushrooms meet, form gametes and fuse. This diploid cell now divides by mitosis to produce the large structure which grows above ground, the mushroom itself. The mushroom consists of a stalk growing from the ground, with a cap, usually in the form of an umbrella. When the cap develops, it produces a series of thin membranes, called gills since they superficially resemble fish gills. Along either side of the gills, the outermost layer of diploid cells undergo meiosis to produce four spores each. Obtain a slide labeled "*Coprinus*". This is a very small mushroom from which thin slices have been cut through the cap, and mounted on the slide. ●a) Label the cross-section as seen under scanning magnification, with the cap, stalk and gills. This should resemble an old-fashioned wagon wheel with the outer "rim" the cap, the center "hub" being the stalk, and the "spokes" being the gills. Then go to higher magnification after centering the slide on the gills. You should be able to see some of the spores which are produced. ●b) Label one of the basidiospores. ●c) Are these spores produced by the gills diploid or haploid? ●d) Does the mushroom represent sexual or asexual reproduction?

Exercise 3

In addition to sexual reproduction, which is always defined on the basis of the union of two gametes, many fungi also exhibit asexual reproduction. Asexual reproduction is a type of cloning. Mitosis, which always produces two genetically identical cells, is used to generate additional copies of the organism. In bread mold, spherical sporangia are produced at the top of vertical hyphae (called sporangiophores). Obtain a slide labeled "*Rhizopus* Sporangia" and observe it under medium magnification. ●a) On the photograph provided in the answer sheets label a sporangium and sporangiophore. When the sporangium is mature, it will have produced dozens of spores by mitosis. The sporangiophore holds up the sporangium so that spores can be scattered by the wind to new locations where more fungi might grow when the spores germinate. Can you see one sporangium which is mature and releasing spores? ●b) Are the spores produced in the sporangium diploid or haploid?

Obtain a slide of *Penicillium*. This is the blue-green colored mold which grows on many citrus fruits and from which the antibiotic penicillin was first isolated. Examine the hyphae of this fungus. Because they are lightly stained, you may wish to decrease the amount of light coming through the diaphragm. Because the hyphae are so thick in the center of the slide, scan around to the edges of the fungal mass. Look for structures which resemble the arm and hand of a skeleton. The structure which resembles the "skeleton arm" is termed a conidiophore, which is equivalent to the sporangiophore of bread mold. At the tip of the sporangiophore, mitotic divisions produce chains of small round conidia (the "fingers" of the skeleton arm). Each conidia represents a spore produced by mitosis. ●c) On the photograph of *Penicillum* label the conidiophore and conidia.

The unicellular fungi, such as yeast, perform asexual reproduction by simply budding off a portion of cell which can then grow to a large mature cell and begin producing buds itself. Obtain a slide labeled "budding yeast" and examine it under high magnification on your microscope. ●d) On the photograph provided, label the mature yeast cell and the bud. ●e) What is the difference genetically between the mature cell and the bud? ●f) Is the bud a small diploid or haploid cell?

Kingdom Plantae

Plants have an interesting life cycle which involves both sexual and asexual reproduction. There are actually two different forms comprising each species. A haploid gametophyte generation which reproduces sexually using gametes, and a diploid sporophyte generation which reproduces asexually using spores. The haploid gametophyte generation begins when a haploid spore germinates and grows mitotically into a multicellular organism. This haploid, multicellular organism is one type of plant which reproduces sexually. Male gametophytes antheridia which form haploid sperm by mitosis. Female gametophytes form archegonia which form haploid eggs by mitosis. When a haploid sperm fertilizes a haploid egg, a diploid zygote results. The zygote is the first cell in the diploid sporophyte generation. The zygote also grows via mitosis into a multi-cellular plant, the sporophyte. The sporophyte will produce some type of sporangium where diploid cells will undergo meiosis to form haploid spores.

Exercise 4

Let us now examine the different stages in the life cycle of a variety of plants. Obtain slides labels "*Minium* antheridia" and another "*Minium* archeogonium". These slides show the gametophyte generation in moss. The gametophyte generation is the dominant or common stage in moss. It always reproduces sexually. In this moss, there are separate male and female gametophytes. The male produces many sperm within each antheridium while the female produces a single egg within each archegonium. On your answer sheet there are photographs of these structures. For these diagrams, indicate ●a) which generation is represented ●b) what type of plant is being examined, and ●c) label an antheridium and an archegonium. ●d) Do the antheridium and archegonium represent male or female reproductive structures?

Next, obtain a slide of the fern prothallus. This represents the gametophyte of the fern. It is a small plant, usually much less than one inch across. The entire prothallus is included on your slide. Notice that in the fern gametophyte, antheridia and archegonia are produced on the same, not different, plants. If the gametophyte is heart-shaped, then archegonia are located near the tip or bottom of the heart while antheria are located near the "cleft". On the photograph on your answer sheet label ●e) the archegonia and antheridia. ●f) Does this represent sexual or asexual reproduction? Finally, obtain the slide marked "mature anther". In the flowering plants, the gametophyte is represented by a small group of cells within the flower, which develop from the spore. The male gametophyte is represented by the pollen grains which develop within the anther. On the slide you can see the hollow chamber within which they develop. ●g) On the photograph on your answer sheet, label the pollen as the male gametophyte. Within the ovary of the flower a similar event occurs by which haploid spores develop into a female gametophyte from which the egg is produced by an archegonium.

Exercise 5

The sporophyte generation represents the other half of the plant life cycle. It begins when the egg is fertilized to produce the diploid zygote. The zygote then grows by mitosis to produce another organism, the multicellular diploid sporophyte. The sporophyte generation always reproduces asexually when certain diploid cells under go meiosis to form haploid spores. Meiosis occurs within a chamber (sporangium) inside the sporophyte. In moss, the sporophyte begins as a long stalk which grows from the top of the female gametophyte, where the egg was fertilized. At the tip of this stalk, the sporangium (called a capsule) develops. Obtain a slide of the "moss capsule" and examine it under scanning magnification. Notice the two chambers within which meiosis produces the numerous spores. ●a) Label the capsule and spores in the photograph. ●b) What generation does this photograph represent? ●c) What type/group of plants is represented? The complete life cycle of green plants is given below, consisting of both gametophyte and sporophyte generations.

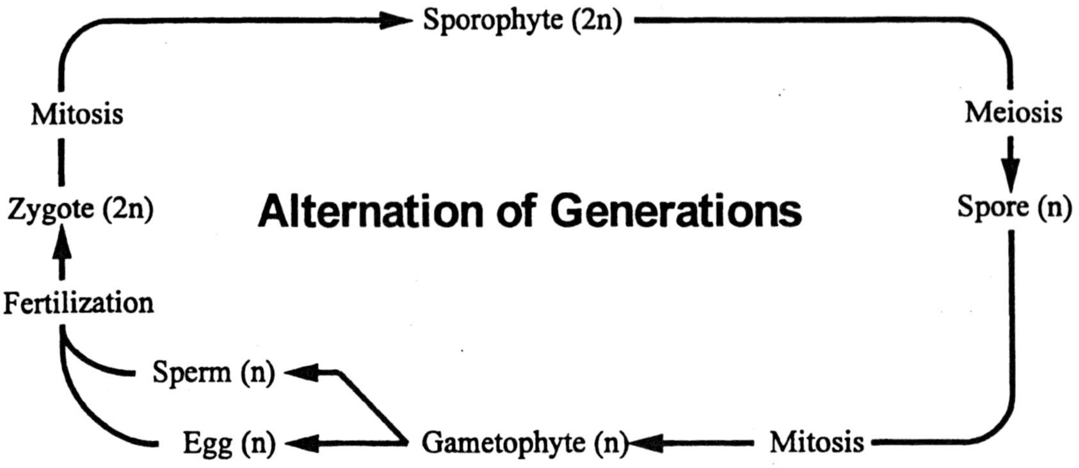

In the ferns and flowering plants, the sporophyte is the dominant or common generation. The typical fern or flowering plant has grown from the zygote. When mature, sporangia develop on the back side of the leaves. Obtain a slide labeled "fern sporangia" and examine it under scanning power. This slide contains a loose arrangement of sporangia scattered across the slide. On the intact leaf, the sporangia are housed under a protective umbrella-like structure called the infundibulum. ●d) on the photograph in your lab, label one of the sporangia and at least one spore within it. ●e) What generation does this photograph represent? ●f) What type/group of plants is represented?

Name ______________________.
Number _____________________.

Kingdoms Fungi and Plantae

Kingdom Fungi

Exercise 1.

a. Label the following structures:

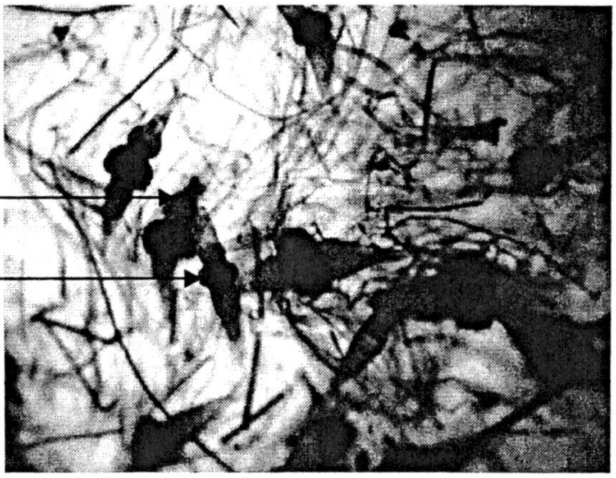

____

____

b. Type of reproduction? ____________________

c. Ploidy of Zygospore? ____________________

d. Ploidy of Gametes? ____________________

Exercise 2.

a. Label the cap, stalk and gills.

____

____

____

____

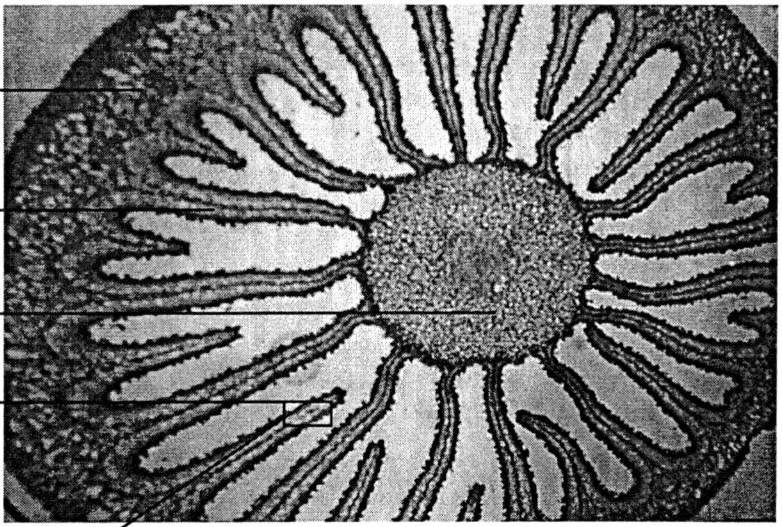

b. ______________________________

c. ______________________________

d. ______________________________

Exercise 3.

a. .Label the sporangium and sporangiophore:

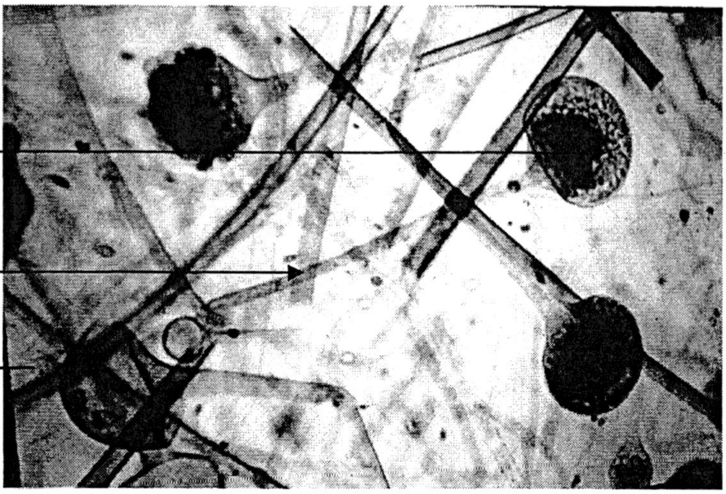

b.

c. Label the following structures:

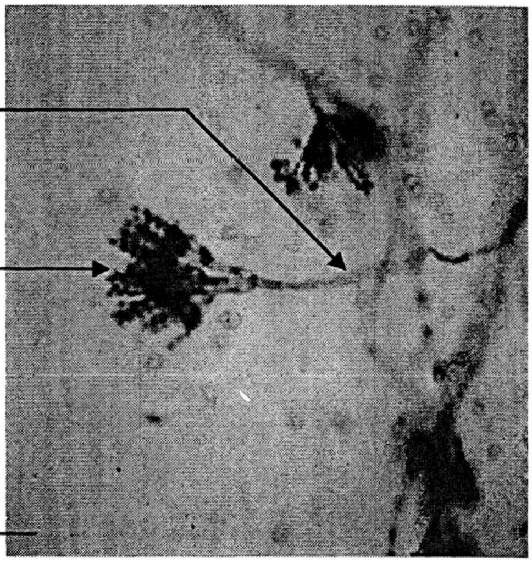

d. Label the yeast cell and a bud.

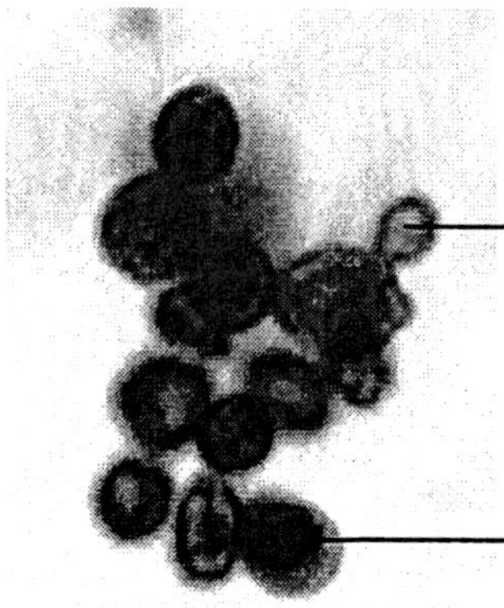

e. ______________________________

f. ______________________________

Kingdom Plantae

Exercise 4

a. What generation is represented below? ______________________

b. What group of plants is represented below? ______________________

c. Label the indicated structures.

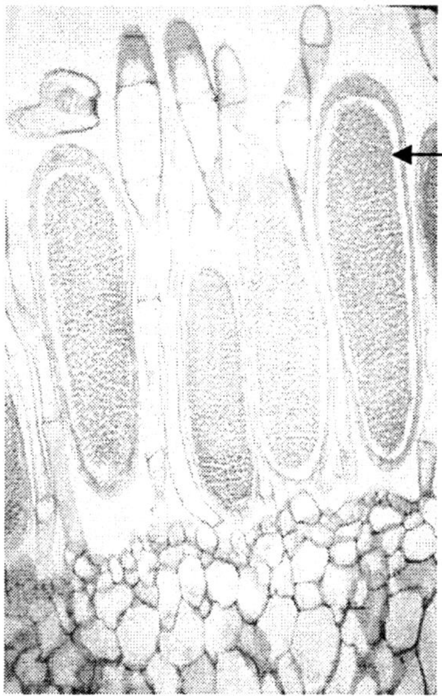

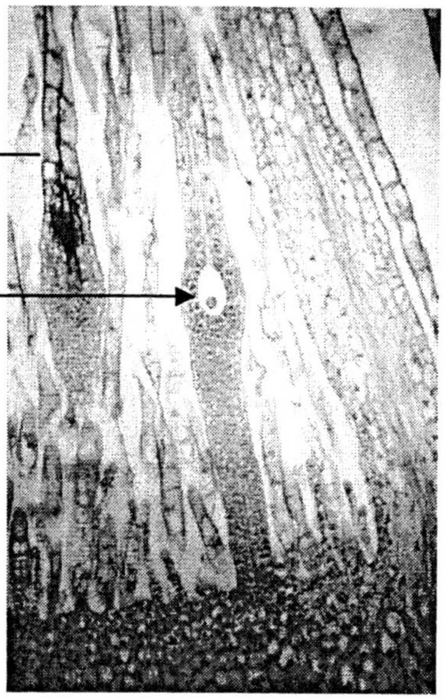

d. Do these structures represent male or female reproductive structures?

______________________ ______________________

e. Label the following structures in the Fern Gametophyte:

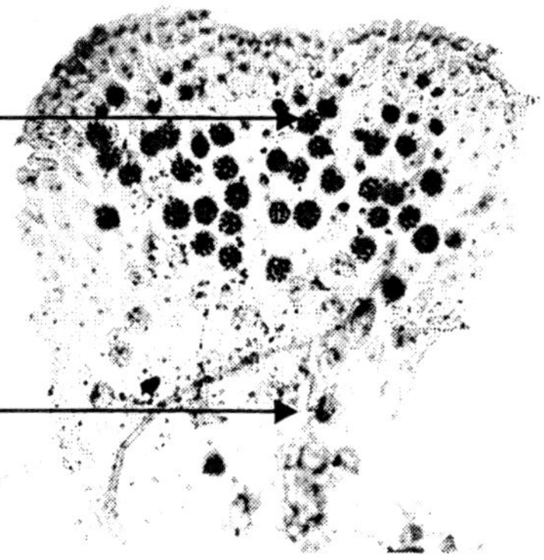

f. Type of reproduction? ______________________

g. Label the pollen as the male gametopyte

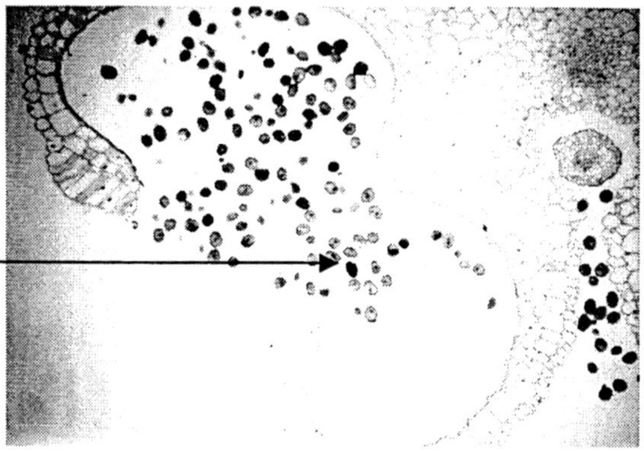

Exercise 5

a. Label the following structures:

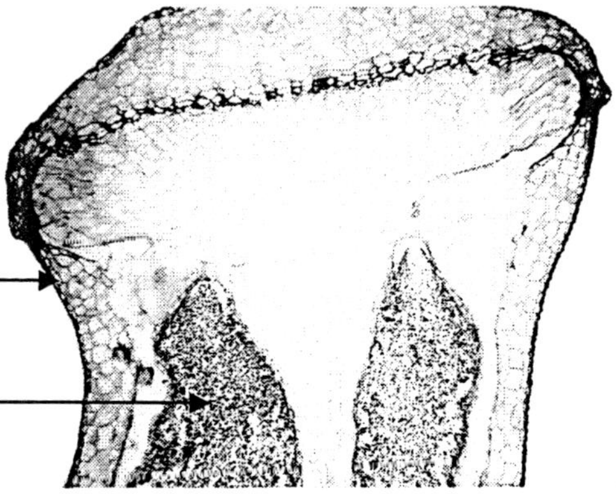

b. Generation?

c. Group of plants?

d. Label the following structures:

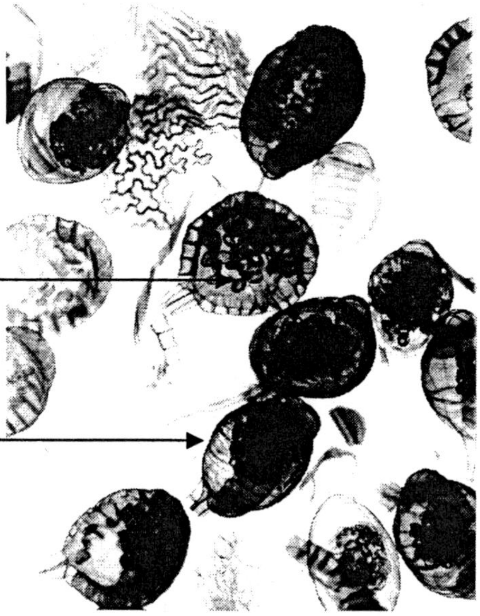

e. Generation?

f. Group of plants represented?

Kingdom Animalia

Introduction

The Animal Kingdom includes multicellular organisms which are comprised of eukaryotic cells, are non-photosynthetic, and do not have cell walls. This is a large diverse group of heterotrophic organisms ranging from the relatively simple sponges to the more complex and familiar animals including humans. It is the purpose of this lab to introduce you to the phyla of the Animal Kingdom and the creatures which they contain.

Objectives

Upon completing your study of this unit, you should be able to:

1. Recognize and be able to place each animal examined into its proper phylum.
2. Describe the general and special characteristics of each phylum.
3. Be familiar with the meaning of the names of each phylum.
4. Describe the general and special characteristics of the classes of Chordates examined.
5. Be able to identify animals in the classes of Chordates examined.

You will be examining the animals of each phylum in order to become familiar with their appearance and their characteristics. Phylum characteristics will be divided into two categories: general and special. The general characteristics that will be examined are common for all phyla, while the special characteristics are those unique attributes that make each phylum different and interesting. The following are the general characteristics that will be presented for each phylum:

1. Symmetry – animal bodies are arranged in a few basic ways:
   a. Asymmetrical (A) – no real organization; unorganized mass.
   b. Radial (R) – arranged with similar structures progressing outward from a center like the spokes of a wheel.
   c. Bilateral (B) – organized with two-sidedness or handedness, e.g., right side vs. left side.
2. Digestive tract.
   a. Complete (C) – have complete tract with mouth and anus.
   b. Incomplete (I) – cavity with only single opening.
   c. Absent (–) – no digestive tract at all.
3. Respiratory system.
   a. None (–) – Have no respiratory system; oxygen and carbon dioxide exchanged by diffusion through outer tissue layers.
   b. Tracheae (T) – extensive branching system of air passages that open along the abdomen and extend throughout body.
   c. Book lungs (B) – opening into a series of flattened, parallel, leaflike air pockets.
   d. Gills (G) – structures over which water circulates; oxygen and carbon dioxide are exchanged by diffusion with tiny blood vessels which permeate gills.

e. Lungs (L) – opening into extensively branched cavity; oxygen and carbon dioxide are exchanged by diffusion with tiny blood vessels which permeate lungs.

4. Circulatory system – pump with vessels for circulating fluid.
 a. None (–) – have none.
 b. Open (O) – pumps fluid into tissues where it trickles around and collects in cavities called sinuses.
 c. Closed (C) – pumps fluid throughout body in tubes or vessels.

General Characteristics of the Phyla of the Animal Kingdom.

Phylum	Symmetry	Digestive Tract	Respiratory System	Circulatory System
Porifera	A or R	–	–	–
Cnidaria	R	I	–	–
Platyhelminthes	B	I	–	–
Nematoda	B	C	–	–
Mollusca	B	C	G or L	O
Annelida	B	C	–	C
Arthropoda	B	C	G, T, or B	O
Echinodermata	R	C	–	O
Chordata	B	C	G or L	C

The special characteristics are of course different for each phylum and will have to be discussed individually. As you progress through your overview exploration of the Animal Kingdom, focus on the general and special characteristics of each group. Try to observe as many of the characteristics as you can. The phylum name itself is derived from one of the important characteristics of the phylum that you should try to observe.

Note: To avoid crowding, you may do the exercises in any order. Relax and enjoy your tour through the Animal Kingdom!

Exercise 1. Phylum Porifera (pore + bearing).

These "pore-bearing" animals are the sponges. The numerous pores extend from the outside of the sponge's body through an extensively branched system of canals into a central cavity called the spongocoel. Flagellated cells called collar cells lining the canals cause water to enter the pores, proceed through the spongocoel, and exit through a larger opening typically on the top of the sponge referred to as the osculum. Food particles and oxygen are obtained as the water circulates through the sponge; wastes are carried out through the osculum. The body of the sponge has an internal structural skeleton comprised of spicules, which are needlelike or fibrous in nature, depending upon the type of sponge. Sponges live in fresh and salt water, but the marine types are larger, more common, and more familiar.

Your lab instructor has placed a sponge under a stereo microscope. Examine this whole sponge. Observe the pores as numerous openings along the sides. Observe the osculum as a larger opening of which there will be only one or a few. Inside this larger opening is the spongocoel. ●a) Draw a picture of a sponge and label pore, osculum, and spongocoel. Show the flow of water through the sponge using arrows.

Exercise 2. Phylum Cnidaria (nettle-like).

The cnidarian organisms are so named because many of their cells, especially on the tentacles, have stinging organelles (nematocysts) which they use to obtain food. Jellyfish, sea anemones, corals, and freshwater hydras are all included in this group. Some exist in a stationary, attached "polyp" form with the tentacles reaching upward (hydra, anemone). Others usually exist in a free-floating form, the "medusa", with the tentacles hanging downward (jellyfish).

Examine preserved specimens of coelenterates (coelenterates is another name for the Phylum Cnidaria) and observe the tentacles. If coral is available, what you will see is not the organism but the secreted "house" of the cnidarian which once lived in the tiny cavities throughout the coral. ●a) Name 3 different kinds of Cnidaria on the answer sheet.

Your lab instructor has placed live hydra with *Daphnia* (these are swimming around in the water and are food for the hydra) under a stereo microscope. You might see a hydra catching a *Daphnia* with its tentacles or you might see a bulge in the hydra's 'stomach'. The bulge would most likely be caused by a recently eaten *Daphnia*. ●b) Draw a hydra and label tentacles.

Exercise 3. Phylum Platyhelminthes (flat + worm).

There are three types of these flat and somewhat wormlike animals including free-living types that inhabit lakes and streams and two parasitic types. Flukes and tapeworms live primarily in the digestive tract of vertebrates (animals with backbones). Examine a live culture of *Planaria* which is one of the free-living flatworms. These are best viewed using a stereo microscope. The free-living flatworms are most common on the undersides of rocks in slow-moving streams. ●a) Draw a picture of a flatworm on the answer sheet.

Examine a prepared slide of a tapeworm. The most common tapeworm in this country is *Taenia*, the beef tapeworm. It attaches to the intestine in humans by hooks and suckers on its head (scolex). If you examine a prepared slide you can see these structures. If you look at the smaller 'head' section you should see suckers and hooks under high power. In many of the larger, square/rectangular sections (proglottids) you will find lots of ball-like structures. These are eggs that could be released to infect a cow. ●b) Draw a tapeworm on the answer sheet and label the scolex (suckers and hooks) and several proglottids.

Eggs are shed in the feces and can be spread to cattle, usually via water or on grass, where they are ingested. The eggs hatch in the cow's intestine into a larval form that burrows into the muscle and encysts. The cow is called the parasite's secondary host since it is the species infected by the larval form. If a human eats uncooked or rare beef containing the cysts, the cyst is freed during digestion and develops into a new, adult tapeworm. The human is called the primary host since it is the species infected by the adult form of the parasite. ●c) Write the name of the primary and secondary hosts for the beef tapeworm.

Exercise 4. Phylum Nematoda (round worms)

The round worms are so named because they are essentially a tube within a tube body design because they have a complete digestive system. Although there are free-living roundworms in soil (nematodes), there are many roundworms that are parasites of humans. The following table lists some of the more common human parasitic roundworms.

Genus (common name)	Where Located In Humans	How Worms Are Transmitted	Problems Caused
Ascaris (roundworm)	Intestine	Eggs in feces	Burrow into body
Trichinella (trichina worm)	Muscle	Eating pork	Burrow into body (trichinosis)
Enterobius (pinworm)	Intestine	Eggs in feces	Rectal itching

Examine a jar with *Ascaris* worms (**DO NOT REMOVE THE WORMS FROM THIS JAR**). The worms with a curled tail are males and the ones with straight tails (generally the larger worms) are females. As you can see not all roundworms are microscopic. Infection by *Ascaris* worms is not uncommon in many countries. If you have a severe infection, it could lead to intestinal blockage which could be fatal. Medication can be taken which can rid the infected person of roundworms. Some of the medications essentially paralyze the worm so that it loses its hold on your intestinal wall. Having lost 'its grip', the worm(s) should come out in your next bowel movement.

Pinworms are much smaller than *Ascaris* worms and are the most common parasitic intestinal roundworm infecting people in the United States (note: since a pinworm looks like a tiny *Ascaris* worm, we will not have a specimen to examine). Up to 16% of the U.S. adult population is infected. The worm lives in your intestine and when it is ready to lay eggs, the female crawls to your anus and deposits eggs in this area. This whole process causes irritation which prompts the infected person to scratch where he/she has an itch. Alas, without proper washing of hands, the infected person, with pinworm eggs on their fingers, may reinfect him/herself. The simplest detection method for pinworm is to use the sticky side of tape and apply it gently to the anal region. If you are infected, the tape will likely pick up some pinworm eggs. These eggs can be seen using a microscope.

●a) Considering how worms are transmitted (see chart above) how could you avoid getting *Ascaris* worms, trichina worms and pinworms.

Exercise 5. Phylum Mollusca (soft).

The "soft-bodied" animals have no internal skeleton, but many of them secrete a protective shell. The group includes the clam, snail, and octopus. The most significant aspect of this phylum as far as humans are concerned is that we eat many of them. .a) Examine representative molluscs and **list four on your answer sheet that are not included in the list above**. Notice the "soft bodies". ●b) Besides their soft bodies, why are such different creatures as snails and clams in the same phylum (see Table on General Characteristics of the Phyla of the Animal Kingdom)?

Exercise 6. Phylum Annelida (little ring).

The bodies of the animals of this phylum are divided into many little segments which externally appear as "little rings" around the wormlike bodies. Earthworms, leeches, and a few marine types called sandworms are the most common types of annelids. Earthworms are extremely important in soil aeration and fertility... they are not bad fishing bait either!

Examine representative annelids and ●a) list the three main types of annelids. ●b) The phylum name 'Annelida' is descriptive of what feature of the animals you examined?

Exercise 7. Phylum Arthropoda (jointed foot).

The arthropods have an exoskeleton (hard outer covering like a suit of armor) made of chitin which has joints in the legs so that they can bend. This is the largest phylum in the Animal Kingdom, and its 800,000 species include 75% of all known animals! The group includes organisms like spiders, insects (flies, cockroaches), and crustaceans (crabs and crayfish). Certainly the impact of the insects on humans is tremendous and obvious. Spiders you may have heard about include the black widows and brown recluse spiders. These are the two groups of spiders in the United States that can give severe to sometimes fatal bites.

●a) Examine representative arthropods, and **list four of them on your answer sheet which are not included in the list above**. Notice the joints in the legs. ●b) The phylum name 'Arthropoda' is descriptive of what feature of the animals you examined?

Exercise 8. Phylum Echinodermata (hedgehog + skin).

The hedgehog is a mammal with sharp lengthy spines, and the echinoderms are also characterized by the presence of spines. The echinoderms are all marine (live in the ocean) and include sea stars (starfish), brittle stars, sand dollars, sea urchins, crinoids, and sea cucumbers. One of the physiologically unique aspects of this group is a water vascular system which takes in water through an opening called the madreporite and moves it through a series of canals leading to the tube feet. The seastar can move with its tube feet located on the undersides of the seastar's rays (arms) by using water pressure to extend or retract the tube feet.

●a) Examine representative echinoderms and name three on your answer sheet. Notice the "spiny skin". Examine a preserved seastar. On the top side you can easily see the spines. On the central disk from which the arms originate, you can see a small, round, hard plate – that's the madreporite, the opening to the water vascular system. On the underside of the seastar, you can see many tiny tube feet extending along a groove that runs along the underside of each ray. ●b) Make a drawing of the seastar and label the rays, madreporite and tube feet.

Exercise 9. Phylum Chordata (cord bearing).

This is the phylum that includes humans and familiar animals like fish, reptiles, birds, frogs, dogs, cows, etc. The phylum name comes from a structure called the notochord, which is a rod-like semirigid supporting device that runs the length of the body during embryological development. Only a few types retain the notochord as an adult. In most adults the longitudinal supporting role is played by the vertebral column or backbone. Animals with backbones are called vertebrates.

We will examine several classes of vertebrates. You should be able to categorize each organism examined below into their respective classes in addition to being able to identify each organism examined. Write down any additional characteristics, beyond those given below, which will help you correctly categorize and identify the organisms examined.

1. Jawless Fish – As the name implies, these fish lack jaws like sharks and bass have. Examine a sea lamprey (a native fish of Virginia). Notice the sucker-like mouth. This particular species attaches to fish and sucks their blood. The blood loss is not usually fatal to the parasitized fish. It is rather an infection in the open wound left after the lamprey detaches that causes death.

2. Cartilaginous Fish – As the name implies, these fish have a cartilaginous skeleton. They lack bones like you and I have and hence their bodies are a bit more 'rubbery' than ours. Examine a shark and a stingray. Most cartilaginous fish are marine. Large rays (manta rays) can have a 'wingspan' of 20 feet. The larger sharks such as the great white shark can grow in excess of 20 feet in length.

3. Bony Fish – These fish have skeletons made of bone. Examine a gar (fish with a long skinny mouth with lots of teeth (a fish-eating fish), a bluegill (a common fish in ponds, notice the black spot at the posterior base of the fin along their back (dorsal fin)), and a freshwater eel (looks like a snake but it has fins). All these fish can be found in Virginia.

4. Amphibians – These organisms typically have smooth to rough skin (no scales as in reptiles), four legs and at least some part of their life history is tied to water. For example, many live part of the life in the water (frog tadpole) and part on land (adult frog). Examine a bullfrog (large green frog), American toad (rough, 'warty' skin), red-spotted newt (has 4 legs, spots, and a flattened tails used for swimming). Note the red-spotted newts has a land-dwelling form which called the red eft. It is bright orange and can often be seen walking around during the day in the woods. The bright color is a warning to predators and indicates the eft does not taste good. All these amphibians are common to the Virginia area.

5. Reptiles – These have scales and eggs with leathery shells which do not need to be laid in water like most amphibian eggs. These include animals such as turtles, alligators and crocodiles, snakes and lizards. Examine the shell of a turtle (note the backbone is attached to the shell), a 5-lined skink (a lizard common in suburban areas, the scales are very small and may be hard to see), a fence lizard (scales should be very easy to see on this lizard), copperhead and timber rattlesnake. The last two snakes are the only two venomous snakes in central Virginia and should be avoided if seen outdoors. All other snakes in central Virginia are non poisonous. The copperhead has hourglass-like markings along its body while the timber rattlesnake is banded and, of course, has a rattle. Both have heat sensitive pits between the eye and nostril that can detect body heat from warm-blooded prey such as mice. All of the reptiles examined are from Virginia.

6. Birds – Feathers are the diagnostic characteristic for this class. They also have lightweight bones and a large breastbone for flight muscle attachment. The shells of bird eggs are hard in comparison to reptile eggs. Birds range in size from hummingbirds (as small as 2 ½ inches and weighing less than some types of moths) to ostriches (to 9 feet in height and 345 lbs.). Some have wingspans in excess of 10 feet. The arctic tern breeds north of the arctic circle and then migrates south to the Antarctic, a trip that exceeds 11,200 miles. Some birds migrate over the Himalayan mountains (29,141 feet). They can fly at high altitudes where oxygen is scarce because of a very efficient respiratory system. Your lab instructor will show you some slides or skins of common birds of Virginia. Be able to identify a robin, cardinal, turkey vulture, starling (an introduced species) and mourning dove.

7. Mammals – Hair and mammary glands (milk producing glands) are diagnostic characteristics of mammals. Skulls often provide clues as to the type of food the species eat. Examine several skulls. The beaver skull has large, brown incisors (incisors are teeth in the front of the jaw which are used to snip or bite) followed by a gap. After the gap are large, somewhat flattened grinding teeth called molars. This tooth arrangement is typical of gnawing animals broadly grouped as rodents. Deer lack incisors on the upper jaws. Their skulls are also fairly long. Carnivores have canine teeth in addition to incisors and molars. Canine teeth are sharp, cone-like teeth just behind the incisors and are used to pierce prey. You should be able to distinguish the skulls of two carnivores. One is a bear (the larger, stout one) and other is a coyote. All these mammals can be found in Virginia.

In addition to being able to categorize each organism examined into their respective class and being able to identify each organism examined, answer the following questions.

●a) How can you distinguish between the following: sea lamprey, freshwater eel, a snake.

●b) If you had a fish that you could dissect, how could you determine if it was a bony or a cartilaginous fish?

●c) People often confuse lizards and salamanders. How could you distinguish between them?

●d) Name one or two key characteristics that allow you to identify a
robin
cardinal
turkey vulture
starling
mourning dove

●e) How could you distinguish a beaver skull from a deer skull, a beaver skull from a carnivore skull?

●f) What if you saw a creature that had webbed feet and a bill like a duck, had hair and provided milk to its offspring which were hatched from eggs. What class would this animal belong to? Do you know the name of this animal which lives in Australia?

Name ____________________________

Number __________________________

Kingdom Animalia

Exercise 1. Porifera.

a)

Exercise 2. Cnidaria.

a)

b)

Exercise 3. Platyhelminthes.

a)

b)

c) primary host ______________ secondary host ______________

Exercise 4. Nematoda.

a) Ascaris

trichina worm

pinworm

Exercise 5. Mollusca.

a)

b)

Exercise 6. Annelida.

a)

b)

Exercise 7. Arthropoda.

a)

b)

Exercise 8. Echinodermata.

a)

b)

Exercise 9. Chordata.

a)

b)

c)

d) robin

cardinal

turkey vulture

starling

mourning dove

e) beaver from deer

beaver from carnivore

f)

Ecology:
A Focus on Communities in Two Contrasting Ecosystems.

Objectives.

1. Describe why tropical rain forests are located primarily near the equator.
2. List 3 characteristics of tropical rain forests.
3. Describe why deserts are located primarily near 30° N and 30° S latitude.
4. List 3 characteristics of deserts.
5. Define ecology, habitat, community and ecosystem.
6. Define keystone species, competition (intra and interspecfic, exploitative and interference), niche partitioning, predation, herbivory, commensalism, mutualism and introduced species. Illustrate each of the terms with examples from the videos.

Imagine yourself arriving in Manaus, Brazil (a city along the Amazon River) for your month-long Amazon ecotourism trip. Your immediate impression upon getting off the plane is that it is hot and humid. Maybe on day 7 of your trip, while you are drifting along the Amazon River, you realize that it had rained every day. Not a shower lasting all day, but maybe one or two hours a day. If you could stay for a couple of months you would discover that the number of hours of daylight remains about the same year round – 12 hours of light to 12 hours of darkness. All of these characteristics of the Amazon region are linked and due to the fairly constant input of solar energy at the equator. This input of energy causes the air masses, laden with water vapor, to rise. This warm air mass cools as it ascends because air pressure decreases as you go up. Since cold air cannot hold as much moisture as warm air, it drops part of its load of water in the form of precipitation. This cycles repeats itself virtually every day in the warm, moist tropical rain forests (TRF) encircling the globe near the equator. The major rainforests are located in the Amazon region of South America, the Congo region of Africa, parts of Southeast Asia, Indonesia and Northern Australia.

If you could follow these air masses that dropped their moisture in the equatorial regions you would find that they continue to rise to the boundary between the troposphere and the stratosphere (about 17 km up). This dry air divides and flows to the north and south as it continues to cool during its journey toward the poles. As the air masses advance they become denser and around 30° N and 30° S latitude the air has become dense enough that it begins to descend. Because air pressure increases as you descend toward the earth, the dry air also heats up. This hot, dry air now sweeps across the surface of the earth and may actually pick up moisture from the surface of the earth. These conditions produce our deserts. The majority of the deserts are located at 30° N and 30° S latitude. In the Northern Hemisphere, these deserts include our Sonora desert, the Sahara desert in Africa, and the deserts of the Middle East. The Southern Hemisphere deserts include areas such as the Namib desert in Africa and the "outback' region of Australia.

The organisms that live in these two ecosystems, the TRF and the desert, experience very different sets of environmental conditions. The TRF ecosystem has fairly stable temperatures averaging 25° C year-round, high humidity of 80% or more, and lots of precipitation fairly evenly distributed throughout the year (it can exceed 200 cm/yr and is rarely less than 130 cm/yr). If you can stand the warm and humid air this is not a bad place to live. In fact, there are more species in the TRF than in all the other ecosystems combined. Typically no single tree species dominates in a TRF but rather there are a large variety of species (high diversity), with up to 500 species of trees in a square kilometer. The trees form a canopy (up to 30-40 m high) and shade out plants below. Much of the life in a TRF is located in the canopy with a rich array of birds, reptiles, amphibians, insects and epiphytic plants, which are plants that grow on (epi) other plants. These organisms interact with each other in a wide variety of relationships including mutualisms, commensalisms, predator-prey interactions, and competitive interactions (see below for definitions). Many of these organisms live out their entire lives high in the canopy. For example, some frogs find all the water they need for drinking and their offspring (tadpoles) trapped in tree cavities or whorls of epiphytic bromeliad plant leaves.

In contrast to the TRF, the desert ecosystem typically has less than 10 cm of rain per year and a high evaporation rate. Temperatures can fluctuate from a high of 36° C during the day to a low of 15° C at night. So during the day you could use some shade and at night, a source of heat would be nice. Recall what the Lord was like for the Israelites while they wandered through the desert for 40 years. At night He was a ____ and during the day He was a ______. The desert is a great place to live if you don't like humidity but be prepared to be thirsty. Water conservation is a must here and plants often have spines, like our cacti, to protect the water contained in their stems from thirsty herbivores. Relative to a TRF, the diversity of life is less in a desert and species like the saguaro cactus may be predominant, exerting a major influence on community dynamics (i.e. be a keystone species). Daily temperatures fluctuate much more in a desert relative to a TRF and this impacts the activity of many desert animals. Rodents, snakes, scorpions, and coyotes are often active at night when temperatures are cooler and the relative humidity is higher. Other animals, like some species of birds and lizards are active during the hot, daytime hours. These species can handle high temperatures through behavioral strategies and/or physiological mechanisms to avoiding overheating. Regardless of the time period you are observing, you will find in this dry ecosystem, as you did in the TRF, a rich assemblage of interrelated organisms.

Here are some definitions helpful for understanding the lab's introduction you have just read and the films that you will see during this lab:

Ecology is the study of the interactions of organisms both with other organisms and with their physical environment (ex. water, soil, temperature, light).

Habitat is the place or physical setting an organism lives in (ex. desert, forest, pond).

A **community** is the group of organisms that live in the same location (ex. trees, worms, mice, woodpeckers, raccoons, etc make up a forest community).

An **ecosystem** is a community plus the physical environment they live in (ex. all the plants and animals in a forest as well as the rocks, dirt, air, and water make up a forest ecosystem).

A **keystone species** is a species that dominates a community of organisms and is very important in determining the dynamics of that community. It might be an important predator, a plant that provides homes for animals, etc.

Organisms can interact with one another in several important ways:

1) **Competition** – two individuals are trying to use the same resource that is limited.
   There are two types of competition based on the type of individual:
   a) **intraspecific** competition is between individuals of the same species.
   b) **interspecific** competition is between individuals of different species.
   There are two types of competition based on how the individuals interact:
   a) **exploitative** competition is when one individual uses up or exploits a resource so other individuals cannot use it (ex. one bird eats up all the available seeds leaving none for others).
   b) **interference** competition is when one individual keeps another individual away from a resource and interferes with the other individual using that resource (ex. a bird uses a hole in a tree for a nest preventing other individuals from nesting there).

Note: **Niche partitioning** is when two species divide up a resource between them (share it) in a way that reduces or avoids competition.

2) **Predation** – one individual (a carnivore) eats an animal.

3) **Herbivory** – one individual (an herbivore) eats a plant.

4) **Mutualism** – two individuals interact in a way that is beneficial or helpful to both of them (ex. a bee pollinating a flower gets nectar, and the flower gets fertilized).

5) **Commensalism** – two individuals interact in a way that one is helped, and the other is neither helped nor harmed (ex. a tree that provides a nest for a bird helps the bird, and is not harmed by the nest).

Introduced species – species that are foreign to an area and have been introduced by man; usually they are harmful to the other species that live there (ex. rabbits were introduced to Australia and have eaten many of the plants). Humans themselves can have a negative impact on ecosystems, but they can also have a positive impact.

You will view portions of two videos. As you view the videos, answer the questions provided for this lab.

Name ______________________

Number ______________________

Ecology: A Focus on Communities in Two Contrasting Ecosystems.

Saguaro: Sentinel of the Desert

1. From the introductory reading, what causes deserts to be located primarily at 30° N and 30° S latitude?

2. List 3 characteristics of deserts.
a.

b.

c.

3. The saguaro cactus is a keystone species of the Sonoran Desert ecosystem. What are some benefits noted throughout the film that the cactus provides to animals in this desert?

4. What are some birds that nest in and on the saguaro and so have a commensalism relationship with the cactus that benefits the birds but does not harm the cactus?

5. How do great-horned owls and bay-winged hawks niche partition with one another and so avoid competing for the same food?

6. When does intraspecific competition (between individuals of the same species) occur in great-horned owls?

7. What is one introduced species or invader into the Sonoran Desert community that is having a harmful effect on some of the birds there? Is its competition for nest holes an example of exploitation or interference competition?

8. How do rattlesnakes deal with cold temperatures, a physical factor encountered in deserts?

9. How do bats have a mutualistic relationship with the saguaro cactus that benefits both?

10. The roadrunner is a predator (carnivore) of ______________________________.

11. How do the following animals deal with the desert's heat or dryness?
Saguaro cactus.

Desert Cottontail Rabbit.

Kangaroo Rat.

Reptiles.

Birds nesting inside the cactus.

12. Name some herbivores that feed on the saguaro's seeds or fruits. Are these different species intraspecific competitors with each other or interspecific competitors? As they eat up seeds and fruit do they compete by exploitation or interference?

Rain Forest video

1. From the introductory reading, how does a constant input of solar energy at the equator cause rain forests to be located there?

2. List 3 characteristics of tropical rain forests.
a.

b.

c.

3. Describe these characteristics of a tropical rainforest
a. impenetrable nature – why did the early explorers think it was impenetrable

b. actual denseness of the undergrowth

c. quality of the soil

d. location of most of the nutrients

4. What do leaf cutting ants do with the leaf pieces they clip off of plants?

Considering the fact that the fungi raised by leaf-cutting ants can grow only in ant nests, what type of relationship do the ants have with the fungi?

5. Observing throughout the video, describe some methods predators use to catch their prey?

6. Observing throughout the video, describe some methods prey use to escape predation?

7. Plants compete for nutrients and for light. How do plants growing on the ground find light?

8. Describe these characteristics of the relationship between the swollen thorn acacia and ants:

a. type of relationship

b. what benefit does the ant get from the plant

c. what benefit does the plant get from the ant

9. Tropical rainforests receive lots of rain. Describe the mechanisms the following organisms use to deal with this large amount of rain:

a. adaptation of leaves

b. wasp behavior to keep nest dry

10. What does an army ant colony use as a temporary home during raids?

11. Considering interactions between plants and herbivorous insects:

a. what have plants developed in response to the wide variety of leaf eating insects?

b. how have insects responded to the plant's response to herbivory?

The Circulatory System

Introduction

The circulatory system includes the heart, blood and blood vessels. These components work together to meet the demands of the body's tissues. The blood is the vehicle that carries the oxygen and other nutrients to the cells while picking up the cellular waste products. The blood vessels serve as the conduit to the tissues and back to the heart, while the heart acts as the pump that is the main driving force that keeps the blood flowing. The microscopic blood vessel, the capillary, is the vital site where the circulatory system meets the needs of the tissues. It is here where the diffusion of chemicals (oxygen, other nutrients and waste products) occurs between the blood and the watery medium surrounding the body's tissues.

Objectives

1. Identify the 4 chambers of the heart and the major vessels associated with each.
2. Explain why the heart is called a double pump, and contrast the systemic circuit with the pulmonary circuit.
3. Trace the flow of blood through the body including the route through the heart. (major vessels only).
4. Explain the operation of the heart valves and the sounds associated with each when they close.
5. Identify red and white blood cells and platelets and distinguish the general roles of each.
6. Identify the layers to an artery and vein distinguishing differences between the two.
7. List and explain risk factors involved in heart disease.

The Blood

Blood is the river of life that flows through us and transports nearly everything that must be transported from one place in the body to another. If a sample of whole blood was taken and spun in a centrifuge it would separate the blood components into 2 portions. The bottom of the container would have the dense portion of blood known as the **formed elements**, while the upper portion would contain the watery medium of blood, the **plasma**. The three formed elements make up about 45% of the blood volume and the plasma constitutes 55% of the blood volume. The formed elements include the **red blood cells** (erythrocytes), **white blood cells** (leukocytes), and **platelets** (thrombocytes).

The **plasma** is the fluid portion of the blood. The plasma is 90% water, but the other 10% contains over 100 different dissolved substances. These substances include nutrients, gases, hormones, wastes, proteins and salts. The composition of the plasma is constantly changing as molecules are removed and added to the plasma by the cells through diffusion.

Exercise 1

Identification and function of the formed elements

The **red blood cells** function to transport oxygen to the tissues. These cells are made in specialized tissue in the end of bones and shortly before being released into the circulatory system they lose their nucleus. As a result, they have a short life span because they cannot repair themselves. The red blood cell is uniquely designed to carry oxygen. These cells are little more than biconcave discs (like a disc that is pinched in the middle on both sides) that are stuffed full of the oxygen transporting molecule hemoglobin. View the blood slide under the microscope. The red blood cells are the most numerous cells in the field of view and stain pink.

●a) Draw a red blood cell

●b) Do you see a nucleus?

●c) Can you see a difference between the color in the center of the cell compared to the color around the edges of the cell?

●d) What does this mean?

The **white blood cells** are the largest of the formed elements and make up only 1% of the total blood volume. The various white blood cells are crucial in the body's defense against disease. The white blood cells have distinct dark purple staining nuclei. The blood contains 5 different types of white blood cells.
Two of the types are relatively rare, but the neutrophil, lymphocyte and monocyte are all relatively common. Look at the blood smear slide and identify some white blood cells in general. The white blood cells are larger than the red blood cells you identified earlier and have a dark staining nucleus. Now try to find a neutrophil. **Neutrophils** are the most numerous of the white blood cells. They have a multi-lobed nucleus and are phagocytic (phage=to eat, cyte=cell). Identify a lymphocyte. The **lymphocyte** is the smallest of the white blood cells. The cell is only slightly larger than the red blood cell with the dark nucleus accounting for nearly the entire cell mass. One type of lymphocyte produces antibodies, important proteins that are involved in immunity. Now look for the largest of the blood cells, the **monocyte**. You may have to scan the slide a little while to find one of these because they are less abundant than the lymphocyte and neutrophil. The monocyte has a large kidney bean shaped nucleus and is known as a macrophage (macro=large, phage=to eat) after it leaves the circulatory system and enters the tissues.

●e) Draw a neutrophil, lymphocyte and monocyte

A cancerous condition of one of the groups of white blood cells is leukemia. The bone marrow is the site for the production of all blood cells. In leukemia the cell division is too rapid producing numerous irregular-shaped, nonfunctional leukocytes. This cell division is often done at the expense of the normal production of the other formed elements produced in the marrow (the red blood cells and platelets) Examine the leukemia slide and contrast it with the normal blood smear.

●f) What are 2 differences that you observed in the leukemia slide?

●g) Symptoms accompanying leukemia often include bleeding or bruising easily, fatigue, and infection. Can you predict why these symptoms are seen?

The **platelets** are cell fragments from large cells that are found in the bone marrow. They are dark staining, irregular shaped bodies interspersed among the blood cells. Platelets house a variety of chemicals that are instrumental in the clotting process that occurs in the plasma when a blood vessel is injured. Although platelets are not cells, they are sometimes called thrombocytes (clotting cells), because of their function.

●h) Are you able to identify any tiny platelets in the blood slide?
●i) Would you consider them to be more or less abundant than the red blood cells?

The Heart

The heart is the muscular organ that sits in the chest cavity and is only a little larger than a fist. This marvelous organ has the job of beating approximately 100,000 times a day, pumping roughly 8,000 liters of blood daily. It does this job tirelessly, taking breaks only between beats.

The heart is a "double-pump" that distributes blood through two different groups of blood vessels. The right side of the heart forces blood through the **pulmonary circuit**. Blood travels through the pulmonary circuit to the lungs and back. The function of the pulmonary circuit is to provide for gas exchange. Oxygen-poor and carbon dioxide rich blood is sent to the lungs where carbon dioxide is released to the lungs while oxygen is loaded from the lungs into the red blood cells. The blood then returns to the heart oxygen-rich.

The thick-walled left side of the heart forces blood through the **systemic circuit**. The systemic circuit carries oxygen-rich blood from the heart through the body tissues and back to the heart. The function of the systemic circuit is to supply functional blood to all body tissues.

Exercise 2

Viewing the Superficial Anatomy of the Heart and Walls of the Heart.

Using appropriate models (or a sheep heart if available) identify the following of the heart. Notice the shape of the heart. The point of the heart is called the **apex**. This is the part of the heart that is nearest the surface of the skin in humans and is the easiest place to feel the heartbeat. The **base** of the heart is the broad portion from which many great vessels emerge. The base of the heart is angled pointing toward the right shoulder. The heart is normally enclosed in a sac within the chest cavity known as the pericardial sac. The pericardial sac has been removed from the hearts you will examine. As you observe the heart, three different layers can be identified. The outer layer that makes up the thin surface of the heart is the **epicardium** (epi=on, cardium=heart). The thick muscular wall of the heart that makes up the majority of the mass of the organ is the **myocardium**. The thin inner lining of the heart that is slick and is continuous with the lining of the blood vessels entering/leaving the heart is the **endocardium**.

Notice the walls of the chambers of the heart. Remembering the functions of the left and right side of the heart,
●a) Why do you think the left side of the heart is much thicker than the right side of the heart?

Knowing that the left side of the heart is thicker than the right is valuable for an orientation to learning the anatomical features in the next section. The right side of the heart is normally relatively thin because it only needs to generate enough energy to force blood to the lungs and back. A frequent heart disorder seen in smokers is cor pulmonale. Cor pulmonale is right-sided heart failure caused by congestion in the lungs. When blood does not flow freely through the pulmonary circuit, the right side of the heart must work much harder. This thickening of the right side is often seen in cadavers or in x-rays.

Internal Heart Anatomy, Chambers, Valves, and Great Vessels

The heart is made up of 4 chambers. The upper chambers are called **atria**. These chambers receive blood from large veins that enter into them. The atria are relatively small and have small ear-like appendages off of them called auricles. The valves on each side of the heart separate the atria from the lower chambers called the **ventricles**. The ventricles are larger than atria, with the left ventricle making up the bulk of the heart (the entire apex of the heart consists of the left ventricle). Ventricles have the job of forcing the blood up and out the large arteries that leave their chambers.

Valves prevent the backflow of blood into the chamber that the blood just exited. The two **atrioventricular valves** (AV valves) separate the atria from the ventricles. The right side of the heart has the tricuspid or right AV valve, and the left side of the heart has the bicuspid (mitral) or left AV valve. The AV valves prevent the backflow of blood into the atria when the ventricles contract. The AV valves are attached to the heart wall by way of **chordae tendinae** (literally heart strings) that attach into the bulging **papillary muscles** of the ventricles. There are 2 great arteries that leave the heart. The **aorta** leaves the **left ventricle** through the center of the heart. The aorta is the largest artery in the body and quickly gives rise to many branches as it makes an arch. All oxygenated blood that is sent to every body organ first passes through the aorta. When the heart contracts it forces blood up into the aorta. When it relaxes blood begins to fall back towards the heart. To prevent much backflow, the **aortic semilunar valve** closes and seals off the passage from the left ventricle. Similarly, the **right ventricle** forces blood into the **pulmonary trunk** that wraps around the front of the heart. The pulmonary trunk will branch into the pulmonary arteries that carry blood to the lungs for re-oxygenation. The pulmonary semilunar valve is the last valve of the heart and prevents blood from flowing back into the right ventricle once it has been pumped into the pulmonary trunk.

The atria are the receiving chambers of the heart. On the right side of the heart, deoxygenated blood enters the **right atrium** by way of the **superior** and **inferior vena cava**. The superior vena cava drains the head, chest and arms, while the inferior vena cava drains the rest of the body.

On the left side of the heart, blood returns from the 2 lungs through four **pulmonary veins** (two from each lung normally). The pulmonary veins are the only veins in the adult that carry oxygenated blood. These veins empty this oxygen-rich blood that will be sent through the systemic circuit into the **left atrium**.

Try to trace the flow of blood first through the right side of the heart, then through the left side of the heart. Here are some hints that should help you.
1. The left side of the heart is thicker than the right side.
2. Veins enter atria and are thin-walled; arteries leave ventricles and are thick-walled.
3. AV valves separate atria from ventricles. You are in an atrium if you are above the AV valve flaps; you are in a ventricle if you are below the flaps.
4. If you examine a sheep heart, use a probe or wooden pointer to help you identify the large vessels and the chambers they enter or leave. For instance, to identify the pulmonary trunk first find the right ventricle, then try to find an opening that you can poke the probe through. At the other end of the probe you will now have the pulmonary trunk.
5. **This is the general pattern through a side of the heart**. Blood enters the atrium through large veins; from the atrium it passes through an atrioventricular valve to the ventricle. The ventricle then forces blood out of the heart through a semilunar valve into a large artery. Of course the veins, valves, chambers and arteries have special names, but this is the general flow.

●b) Can you find the 3 cusps of the right AV Valve on the model?

From the right ventricle, blood goes out the pulmonary trunk. The pulmonary trunk leaves the heart at an angle near the surface of the heart. Angle a probe from the right ventricle through the pulmonary trunk. Along this length you will find the pulmonary semilunar valve. From the pulmonary trunk blood is sent to the lungs where it will become oxygenated and then return to the left side of the heart.

●c) How many flaps are there in this sort of valve?
●d) Is the pulmonary trunk an artery or a vein?

Locate the left side of the heart. Can you find the veins returning blood from the lungs to the left atrium? These small veins are the pulmonary veins. Normally there are 4 pulmonary veins, but sometimes it is difficult to observe all of them. After entering the left atrium the blood would then travel through the bicuspid valve (left AV valve) to the left ventricle. The large vessel leaving the left ventricle is the aorta. The aorta passes straight up through the center of the heart. If you guide a probe through the aorta you may be able to see the first large vessel branching off of the aorta (the brachiocephalic).

The brachiocephalic supplies the right arm and right side of the head with oxygenated blood.

●e) Can you identify the aortic semilunar valve as you look from the left ventricle into the aorta?
●f) Perform the end of the chapter exercise of tracing blood through the heart by filling in the appropriate blanks.
●g) Your instructor or a lab assistant may choose to point to a chamber of the heart or a blood vessel entering or leaving the heart as a check off to make sure that you understand the heart anatomy before leaving the lab.

Exercise 3

Heart Sounds

The heart sounds are important tools in determining whether the valves of the heart are working properly. The valves are structured to act passively and insure that blood flows in only one direction. If the valves are not working properly, blood may regurgitate into the chamber it just left. This is inefficient and causes the heart to work much harder.

Two distinct hearts sounds can be heard using a stethoscope. The sounds, commonly referred to as "**lubb**" and "**dupp**" are caused as blood forces valves closed. The first sound, lubb is produced as the ventricle contracts and forces blood into the great arteries leaving the heart. As the ventricles contract pressure increases in the chamber. This increase in pressure closes the AV valves, like the slamming of a door, preventing blood from back flowing in the atria. When the heart is at rest the pressure in the ventricle decreases. If not for the semilunar valves this decrease in pressure would allow blood that has just been pumped into the aorta or pulmonary trunk to "fall" back into the ventricle.

Mitral valve prolapse is a common valve disorder (nearly 10% of the population, so it is possible someone in your lab has it). In mitral valve prolapse the valve cusps goes beyond the "closed" position and hyperextend slightly into the atria allowing the regurgitation of blood into the left atrium. This causes a hissing sound when listening to the heart sound with a stethoscope. Any abnormal heart sound is called a **murmur**.

1. Obtain a stethoscope and clean the earpieces with an alcohol swab. When placing the earpieces in the ears they should be angled forward into the ears.

2. You may perform the following exercise on yourself. Place the diaphragm of the stethoscope roughly 2 inches to the left of the bottom of the sternum (breastbone). This is the location where you can feel the pulse the best because the apex of the heart is nearby. Listen to the "lubb" and "dupp" sounds for a minute.

●a) Which of the sounds is longer and louder?
●b) Which of the sounds is short and sharp?

3. When the stethoscope is placed over the apex of the heart you are hearing the closing of the left AV valve the loudest. To differentiate individual valve sounds more precisely, listen to the heart sounds at these locations: 1) between the 2nd and 3rd rib to the right of the sternum (aortic semilunar valve), 2) between the 2nd and 3rd rib to the left of the sternum (pulmonary semilunar valve) and 3) between the 5th and 6th rib to the right of the sternum (right AV valve).

●c) Contrast the sounds you heard when listening to the semilunar valves, with those heard while listening to the atrioventricular valves. (I.e. were the sounds softer or louder, could you hear the lubb or the dupp better in the semilunar valves?).
●d) Could you detect a heart murmur?

The Blood Vessels

The blood vessels make up a closed transport system from the heart to the lungs and body and back again. Blood is forced from the heart in large **arteries** that branch many times into smaller and smaller vessels called **arterioles** and finally into the microscopic **capillaries**. Tiny openings in the capillaries allow small molecules to diffuse between the tissues and the blood. Only small molecules like oxygen, carbon dioxide, urea, glucose, and amino acids can normally pass through the openings. Cells are too big and remain within the blood vessel system. Blood is returned to the heart from the capillaries as millions of capillaries link to tiny **venules,** (miniature veins) that continue to converge to become larger and larger **veins.** These veins merge to form larger veins until they eventually form the superior and inferior vena cava returning blood back to the heart.

Exercise 4

Microscopic Structure of Blood Vessels

Arteries and veins normally travel alongside of one another in the body. Arteries carry blood away from the heart and veins toward it. The heart creates the blood pressure that is the driving force for the movement of the blood through the vessels. To take advantage of the pressure created by the heart, arteries have walls that are able to stretch and recoil to assist in forcing the blood along (this is known as the pulse and generally corresponds with the time that the heart beats and then rests). Veins return blood to the heart and are low pressure vessels (by the time blood returns to the heart the pressure is just a couple of mm Hg compared with an average of 120 mm Hg when it is initially forced from the heart into the aorta). Because veins do not need to withstand the degree of pressure that arteries do, they can have smaller walls. Veins specialize in acting as a blood reservoir for the body. At rest 60% of the blood is found in the venous system. How is it that the *veins* hold more blood than the heart, arteries and capillaries combined? They *have a very large lumen.* (A **lumen** is a passageway, in this case for blood to flow through).

Obtain a slide of the cross section of an artery and vein. Arteries and veins are made up of three layers or tunics. The innermost layer is the **tunica intima** or **endothelium**; this will be very difficult to identify in the microscope because it is only a single layer of cells thick. This layer is continuous through all the blood vessels and even the internal lining of the heart. Capillaries are made up of only a layer of endothelium; they lack the following 2 layers.

The **tunica media** is the middle layer that is more bulky and is made up of the smooth muscle that controls the diameter of the blood vessel lumen. It *will be much thicker in arteries* than in veins and may even contain some elastic fibers near the lumen that look like black squiggly strands. These elastic fibers are important in allowing the arteries to expand and recoil. When you take someone's pulse, this is what you are feeling.

The outermost layer is called the **tunica externa** or adventitia. It is made up of a tough fibrous tissue that supports and protects the blood vessel.

●a) After viewing the slides make a drawing of the artery and vein. The vein sometimes collapses when the slide is made, but the strong artery normally retains its cylindrical shape. Label the following items on your drawing: artery, vein, lumen, tunica intima, tunica media and tunica externa.

●b) Why is the tunica media thicker in arteries?

●c) Which blood vessel has a larger lumen? What is a function of the large lumens (in all of the combined veins)?

●d) What are you feeling when you feel someone's pulse?

Exercise 5

Blood Pressure

As blood is forced out of the heart into the arteries it creates a force against the walls of the blood vessels. This force is known as the **blood pressure**. Because the heart alternately contracts and relaxes, the blood pressure in the arteries varies with this rhythm. When the heart contracts, it forces the blood into the arteries with great pressure. The pressure that corresponds to the contraction phase of the heartbeat is known as the **systolic pressure**. When the heart relaxes, the blood continues to flow through the arteries due to the elastic recoil of the arteries. However, the pressure is not as great during this period. The pressure that corresponds to the period when the heart is at rest is called the **diastolic pressure.** Measuring blood pressure is an indicator on how hard the heart must work to force blood into the great vessels. In a normal individual the heart exerts 120 mmHg of pressure to drive the blood out of the left ventricle into the aorta. By the time blood returns to the heart through the large veins of the systemic circuit the pressure is only a couple of mmHg. If blood vessels are narrowed by atherosclerosis, this can be a potential cause for high blood pressure. Higher pressures in the arteries create a lot of resistance for the heart to overcome to force the blood into them. This increased workload over a long time may eventually cause an enlargement of the heart and damage to the heart muscle. Blood pressure varies greatly between individuals. Average blood pressure is usually 120/80 (blood pressure is written as systolic/diastolic). Hypotension or low blood pressure is not normally a problem. Many petite women have hypotension. A systolic pressure greater than 140 mmHg or a diastolic pressure greater than 110 mmHg usually indicates hypertension.

The sphygmomanometer is an instrument used to obtain blood pressure readings. It is made of an inflatable cuff and pressure gauge. Follow the steps below to determine the blood pressure of your lab partner.

1. Obtain a stethoscope and sphygmomanometer. Clean the earpiece of the stethoscope with an alcohol swab. Roll the sleeve up exposing the arm.
2. Try to identify a pulse around the inner surface of the elbow. Place the diaphragm of the stethoscope here. Now attach the cuff around the arm above the stethoscope diaphragm.
3. As you inflate the cuff to 180 mmHg, you are creating an external pressure that will eventually be greater than the pressure driving blood through the artery in the arm. At this point no blood will be passing into the forearm so don't leave the cuff inflated too long!
4. Turn the valve on the sphygmomanomer to gradually release the pressure in the cuff. Aim for a drop in pressure at a rate of 3-4 mmHg per second. As the pressure falls, watch the pressure gauge and listen for the first soft thudding sounds as blood re-enters the occluded artery. Mentally make a note of this point; this pressure is the systolic pressure.
5. Continue to release the cuff pressure. You will notice more sounds briefly, and then they will stop. The point where the sounds stop will correspond to the diastolic pressure.
6. If you need to repeat the procedure to get another reading, be sure to completely deflate the cuff. If the cuff is not deflated it will give inaccurate, high readings.

●a) Record your blood pressure.

Exercise 6

Risk Factors, Heart Disease and the Schneider Physical Fitness Test

Heart disease is a leading cause of death in the United States each year. **Heart disease** is a broad term for any disorder that affects the blood vessels or muscle of the heart. Atherosclerosis is a form of heart disease that can lead to heart attack (myocardial infarction). In atherosclerosis, fat, cholesterol and other substances collect in the walls of an artery. If the deposits collect in the walls of an artery that supplies the heart they may eventually block the flow of blood to the heart muscle. If flow the flow is not regained, all the cells beyond the blockage die and are replaced with scar tissue.

There are many **risk factors** that increase the likelihood of heart disease. Risk factors are traits that may not "cause" a disease, but are statistically associated with a greater chance of developing a disease. The major risks factors in heart disease are:

1. High cholesterol
2. High blood pressure
3. Cigarette smoking
4. Obesity
5. Stress
6. Lack of regular exercise
7. Diabetes mellitus
8. Family history of heart disease
9. Male Gender

Notice that the first six of these risk factors can be affected by an individual's lifestyle.

The Schneider Fitness Test is a good test to evaluate cardiovascular fitness. The heart rate is measured before and after exercise in this test. Good fitness is indicated by a slow increase in heart rate and a rapid return to the resting rate. Turn to the laboratory report section to perform the exercise.

The Circulatory System

Exercise 1. **NAME___________________ Sect._____**

a. Draw a red blood cell

b. __________________

c. __________________

d. __________________

e. Draw a neutrophil, lymphocyte, and monocyte

f. __________________

g.

h. __________________

i. __________________

Exercise 2

a. __________________

b. __________________

c. __________________

d. __________________

e. __________________

f. Complete the chart with appropriate blood vessels or valves

Right atrium through the tricuspid valve to the ___________________________ through the ____________________ to the pulmonary trunk to the ________________ to the four __________________ ___________ into the __________________ of the heart through the _____________ valve to the _________ __________________ through the __________________ ____________ valve to the aorta to the body capillaries returning to the heart through the superior and inferior __________ _______ to the ___________ ________

Exercise 3

a. _______________

b. _______________

c. _______________

d. _______________

Exercise 4

a. Drawing of an artery and vein. Label the 3 tunics, artery, vein and lumen

b. _______________

c. _______________

d. _______________

Exercise 5

a. What was your blood pressure?

Exercise 6

Schneider Physical Fitness Test

The Respiratory System

Introduction

A person can live weeks without food, days without water, but only a few minutes without oxygen. The respiratory system works together with the cardiovascular system to support cellular activity by supplying cells with oxygen and removing their waste products, which include carbon dioxide. This exercise will examine the two processes of the respiratory system only. The role of the respiratory system is to supply the blood with oxygen and to remove the carbon dioxide from it. This is done through a two-step process: **Pulmonary ventilation** (breathing) is the exchange of air between the atmosphere and the microscopic sacs in the lungs called alveoli. This process involves inspiration and expiration. The second process, **external respiration** is the exchange of gases between the lungs (alveoli) and blood. This process involves the diffusion of oxygen from the alveoli into the blood, and the movement of carbon dioxide in the opposite direction, from the blood to the alveoli.

Objectives

1. Identify the respiratory passages from the nose and mouth to the alveoli and describe functions for the structures.
2. Relate the structures of the ciliated cells and the squamous cells to their functions.
3. Distinguish at least 3 differences between a smoker's lung and a nonsmoker's lung.
4. Describe the relationship between pressure and volume and how this produces inspiration or expiration.
5. From the bell's jar, describe the analogous human structures.
6. Describe the variables involved in determining the pulmonary ventilation rate.
7. Identify the primary stimulus that causes a person to breathe.
8. Describe the chemical reaction that results in a pH change in body fluids.

Exercise 1. Anatomy and Physiology of the Respiratory Passageways

Anatomy and physiology is the study of structure and function. In these exercises you will identify the parts (the anatomy) of the respiratory passages and relate to these parts the associated functions (physiology). The passages of the respiratory system may be divided into 2 regions. The large passageways for air are called conducting zone structures. The small, microscopic passages where gas exchange to the blood is made are called respiratory zone structures.

A. The Conducting Zone

Using appropriate models, charts and preserved specimens of the respiratory system identify the following parts labeled in bold and describe their associated function(s).

Air may enter the respiratory system through the nose or the mouth. There are advantages to breathing through the nose. As air passes into the nose, it enters into the nasal cavity. The **nasal cavity** performs 3 functions better than the **oral cavity**. It warms, filters and humidifies the air so that it is in a near optimum condition to enter the more delicate lower respiratory passages.

The **pharynx,** commonly known as the throat, is the link between the nasal cavity above, the oral cavity in front and the larynx below. The pharynx contains the tonsils. These are patches of cells that play a role in defense against foreign invaders under normal conditions. Because of their location and exposure to the external environment they are prone to infection and it may prove necessary to remove them in a "tonsillectomy".

The **larynx**, or "voicebox", is a cartilage chamber 4 cm (1.5 in) long. Identify the large, leaf-like piece of cartilage called the epiglottis. The **epiglottis** is sometimes called the guardian of the airway. This piece of cartilage keeps food and water from going down the "wrong pipe" (the larynx), directing it into the proper tube, the esophagus. Any time food or fluid enters the larynx it initiates the cough reflex, which "hopefully" ejects the particles from the respiratory passage to prevent damage. To understand how the epiglottis guards the opening in the larynx called the glottis perform the following exercise after viewing the larynx model. Place your fingers around the "Adam's Apple" on your neck, which is part of the larynx. Swallow or take a drink of water and notice what happens to your larynx.

•a) Does your larynx move or remain stationary?
•b) What is the main chamber of the larynx pressing against to prevent your saliva from entering it?

Other functions of the larynx include phonation and sphincter functions. The larynx contains the vocal cords that, when vibrated, produce sounds or **phonation**. These sounds are modified by the muscles of the mouth and resonating chambers in the nasal cavity resulting in voice production. The vocal cords may be closed to act as a sphincter to prevent air from passing through the glottis. This can be done to generate pressure in the abdominal cavity for various reasons.

•c) Talk to your neighbor and try to determine whether the vocal cords vibrate when you inspire (breathe in) or exhale (breathe out). What is your conclusion?

The **trachea**, or "windpipe" is a tube about 12 cm (4.5 in) long that connects the larynx with the bronchi found in the thoracic cavity. The larynx has 16-20 C-shaped cartilage rings that keep this passageway rigid. Many of these rings can be felt in the neck. The trachea has a tissue lining that is common in the large respiratory passages. The lining is made of **pseudostratified ciliated epithelium.** This tissue is important in trapping small particles and moving them towards the throat where they can be swallowed rather than allowing them to damage the delicate tissues of the lungs. Take a slide labeled pseudostratified ciliated epithelium to your microscope. This is a section through the trachea. Observe the following structures that play important roles in trapping and removing particles.

•d) Find the pseudostratified ciliated epithelium and draw a group of cells on high power. These cells can be found next to the free surface in the slide (clear area at the end).

The pseudostratified ciliated epithelial cells contain **goblet cells** that stain light pink-purple. These cells produce the mucous that coats the hair-like cilia seen on the surface of the cells. The mucous is effective at trapping particles while the cilia are constantly in motion acting like oars that propel the mucous toward the throat where it will be swallowed. Each day you produce and swallow over a pint of mucus without noticing it. Other structures that can be seen in this slide include rings of clear cells that add to the mucus production of the goblet cells. The large purple-staining areas in the slide are hyaline cartilage that makes up the cartilage rings in the trachea.

The primary **bronchi** are the 2 large tubes that split off of the trachea heading toward the lungs. From these structures the bronchi branch over 20 times, each branch having a smaller diameter than the one it came from. Once the diameter of the passageways becomes 1 mm, the passages are called **bronchioles**. The smallest bronchioles of the conducting zone are called the **terminal bronchioles**. These lead into the **respiratory bronchioles** of the respiratory zone. The bronchi and bronchioles make up what is collectively known as the bronchial tree.

While air flows freely through most of the respiratory passages, the mid-sized bronchioles represent a potential source of resistance to air flow. The mid-sized bronchioles have a relatively thick muscular lining that may be triggered to contract, which reduces the diameter of the passageway in an event called bronchioconstriction. Bronchioconstriction is the most dangerous symptom seen in an "asthma attack", which may be triggered by an allergen that is breathed in or by other environmental factors.

B. The Respiratory Zone

The respiratory zone structures make up the end of the air passageways. The respiratory zone consists of **respiratory bronchioles** that lead into **alveolar ducts**. The ducts terminate in irregular shaped spaces called **alveolar sacs**. These alveolar ducts and sacs have buds off of them called **alveoli**. The alveoli resemble grapes on a cluster (the alveolar sac). The structures of the respiratory zone are lined with the thinnest of all the possible tissue linings...simple squamous epithelium. The lining of the respiratory zone is in contact with the lining of microscopic blood vessels called capillaries. The site where the lining of an alveolus meets the thin lining of a capillary makes up the respiratory membrane. The respiratory membrane is the essential site for the diffusion of gas.

•e) From the lung slide draw an alveolar duct and an alveolus.
•f) How does the structure of the lung tissue lining fit its function (i.e. why is it good to have a thin membrane forming the lining of the respiratory zone structures)?

Exercise 2. Contrast a smoker's lung with a normal lung

Smoking has many adverse affects on respiratory structures. The hot smoke dries out the tissues of the mouth, larynx, trachea and lungs. The cilia-lined respiratory passageways are damaged and do not function properly. The smoke contains many chemicals that breech the damaged filtering system and enter the lungs. Once in the lungs these chemicals may cause cancer, irritate the tissue and cause fibrosis (scarring) or lead to rupture of alveoli and loss of surface area for gas exchange. The mucous that is normally propelled toward the throat and swallowed travels down to the lungs and must be coughed out or it will cause problems. The chronic cough seen in smokers is known as chronic bronchitis. If the mucous is not expelled it may be used as a nutrient source for bacteria and cause lung infections. It may also act as an extra barrier that gas must travel across making diffusion less efficient.

Observe the microscope slide of smoker's lung (the slide may be labeled *lung with carbon particles*).

Can you identify any of the following traits that may be present in smoker's lung that are not normally seen in a healthy lung? 1) carbon deposits, 2) regions of fibrosis (abnormal thickening in some regions), 3) tumors (abnormal growths that may be caused by the cancer-causing agents in smoke), or 4) vast openings in the lung where little or no tissue is present due to the collapse of respiratory zone passages (this is a trait seen in emphysema).

Contrast the microscope slides of the lung that you viewed earlier with the slide of a lung with carbon particles.

a) Which of the 4 abnormal conditions above could be identified on smoker's lungs you observed? Answers may vary.

Mechanics of Breathing

Pulmonary ventilation, or breathing, is the flow of air into and out of the lungs. **Inspiration** (the flow of air into the lungs) and **expiration** (the flow of air out of the lungs) occur because of pressure changes in the thoracic cavity. Gasses always move from areas of higher pressure to areas of lower pressure. **Boyle's law** is a simple law of physics that states that *the pressure of a given quantity of gas is inversely proportional to its volume* (assuming a constant temperature). This means that when the volume in a container increases, the pressure inside decreases, and vice versa. The body changes the pressure of the gasses in the lungs by changing the size of the thoracic cavity (changing the volume). Volume changes lead to pressure changes, which lead to the movement of gas. Which way the air flows is determined by making the pressure inside the lungs greater or less than the pressure of the gasses in the atmosphere. If pressure is greater in the lungs than the atmosphere exhalation occurs. If pressure is greater in the atmosphere than it is in the lungs inhalation occurs.

Exercise 3. Volume changes, pressure changes and the flow of air.

Using the tape measure with a same gender partner perform the following exercise.
•a) Record the diameter of the thoracic cavity after an exhalation.
•b) Record the diameter of the thoracic cavity after inhalation.
•c) What happened to the volume in the thoracic cavity?
•d) What happened to the pressure in the thoracic relative to the atmosphere?
•e) Which way did the air flow?

Simulate breathing using the Bell's jar lung demonstration apparatus. This device uses a jar to represent the thoracic cavity, a rubber dam as the diaphragm (the floor of the thoracic cavity), a Y-shaped tube as the trachea and bronchial tubes, and two balloons as lungs. Manipulation of the diaphragm changes the pressure inside the balloons causing them to inflate and deflate. What do the following structures represent? •f) The balloons, •g) the jar, •h) the rubber dam, •i) the Y-shaped tube.
•j) When the rubber dam is pulled what happens to the pressure inside the balloons?
•k) What happens to the balloons? •l) This is representative of inspiration or expiration?

Factors influencing the rate and depth of breathing

There are three factors that affect how often and how deeply a person breathes. The stimuli from the most powerful to the least powerful are hydrogen ion (pH), CO_2, and O_2 concentrations in the body fluids. There are receptors in large arteries that monitor carbon dioxide and oxygen levels in the blood. These receptors send impulses to the brain that may alter rate and depth of breathing, but the major stimulus to breathing is the pH of the fluid that bathes the brain. The fluid that bathes the brain is known as cerebrospinal fluid (CSF). This fluid does not have nearly as many buffers in it as the blood has, so small releases of hydrogen ions can alter the pH very quickly. As the pH drops the rate of breathing increases as well as the depth. This can be observed in the following exercise.

Exercise 4. Breath holding, hyperventilation and bag-breathing

A chemical reaction that occurs frequently in the body as tissues produce carbon dioxide into their watery environment is: Carbon dioxide plus water forms carbonic acid which may dissociate to release hydrogen ions and bicarbonate.

The equation is $CO_2 + H_20 \rightarrow H_2CO_3 \rightarrow H^+ + HCO_3^-$

Recall that when H^+ ions are released the pH of a solution becomes more acidic. Phenol red is an indicator that changes from red in a neutral solution to pink→orange→yellow in an acidic solution. Add a few drops of phenol red to a glass of water.

•a) What color is the solution?
•b) What is the pH of the solution? Using a straw blow some bubbles in the water.
•c) What color is the solution now?
•d) What gas did you blow into the water that initiated a chemical reaction with water?
•e) Is the pH of the solution more acidic than it was originally?

When carbon dioxide levels rise in the body fluids, some of the CO_2 moves into the cerebrospinal fluid and the chemical reaction you just performed occurs. This change in pH of the cerebrospinal fluid has a dramatic impact on the rate and depth of breathing.

Working with a partner perform the following exercises using a watch. Have 1 person be the guinea pig for the exercise with questions f-j and another person for exercises k-m.

•f) Hold your breath as long as you can and record the time.
•g) What molecule produced the "drive" to breathe? Now, hyperventilate for 1.5 minutes (take a very deep breath every 3 seconds during this time). When you hyperventilate you are dramatically lowering the CO_2 concentrations in the body fluids.
•h) Now hold your breath for a second time and record the time
•i) Was there a difference between the first and second trial?
•j) Explain.

Observe your partner's rate and depth of breathing during some normal respirations

•k) Record whether they are shallow or deep, frequent or slow. Now have the person breathe into a paper bag for 3 minutes. When a person breathes into a bag he is breathing air that has more carbon dioxide in it than normal and less oxygen. This stale air, when re-breathed, increases the carbon dioxide levels in the blood and eventually lowers the pH of the CSF.

•l) What has happened to the rate and depth of breathing at the end of the 3 minutes (note the rate may not always change but the depth should).
•m) Why did it change?

The Respiratory System

Exercise 1. Anatomy and Physiology of the Respiratory Passageways

Trace the pathway that a molecule of oxygen might travel from the nose to the alveoli of the lungs

Nose or mouth → ________________→ ________________→

________________→ bronchi to

Smaller branches and eventually terminal and respiratory bronchioles →

________________→

________________→ alveoli

Briefly state a function for each of the following structures

Nose:

Pharynx:

Larynx:

Trachea:

Alveoli

a.

b.

c.

d.

e.

f.

Exercise 2. Contrast a smoker's lung with a normal lung

What differences could you observe under the microscope? Describe the appearance of each.

Exercise 3. Volume changes, pressure changes and the flow of air.

a. Diameter of thoracic cavity after an exhalation

b. Diameter of thoracic cavity after an inhalation

c. What happened to the volume as a result of the diameter change

d. What happened to the pressure inside the thoracic cavity

e. Which way did the gas flow

The following parts on the bell's jar are analogous to which structures in the human

f. Balloons

g. Jar

h. Rubber dam

i. Y-shaped tube leading to the balloon

j. What happens to the pressure inside the jar when the rubber dam is pulled

k. What happens to the balloons

l. This is representative of inspiration or expiration?

Exercise 4 Breath-holding, hyperventilating and bag-breathing

a. Initial color of the solution

b. Is the pH acidic, neutral or basic

c. New color of the solution

d. What is the gas that was added to the water that takes part in a chemical reaction with water

e. pH of solution

f. Length of time breath was held in 1st trial

g. Molecule responsible for the "drive" to breathe

h. Length of time breath was held after hyperventilating

i. Was there a difference?

j. Explain

k. Rate and depth of breathing under normal conditions

l. Rate and depth of breathing at the end of bag-breathing

m. Explain the cause for the difference

Special Senses

Introduction

The five special senses of smell, taste, hearing, equilibrium and vision are all located in the head region. The special senses rely on receptors to transform various types of information about the external environment (stimuli) into a language that the brain can understand...action potentials. Action potentials are currents that travel along the cell membrane of neurons. When an action potential reaches the end of the neuron it causes the release of a chemical messenger (neurotransmitter), which can have an effect on cells in the brain. The brain interprets the input and conscious awareness of the stimulus occurs.

The receptors of the special senses are specialized to detect a specific type of stimulus. Chemoreceptors are found in the nose and the mouth; they detect various types of chemicals in solution and are responsible for our senses of olfaction (smell) and gustation (taste). Photoreceptors detect light and are found in the eye for vision. Hearing and equilibrium are dependent upon mechanoreceptors in the inner ear. Changes in pressure or physical contact cause mechanoreceptors to produce action potentials.

Objectives

1. To identify the structural components of the eye when provided with a model, an appropriate diagram, or a preserved sheep or cow eye, and list the functions of each.
2. Define the following terms: *refraction, accommodation, astigmatism, glaucoma*
3. Describe the purpose of the following vision tests: blind spot, visual acuity, and astigmatism.
4. Describe the location and size of the area for the olfactory receptors
5. Describe the structure and function of the taste receptors.
6. Name the 4 basic qualities of taste and point out on a diagram of the tongue the predominant location for sweet, sour, salty and bitter.
7. Name other factors besides taste that influence taste appreciation of foods.
8. Identify the anatomical structures of the outer, middle and inner ear by appropriately labeling a diagram. Describe functions for the structures.
9. Describe the Rinne, Weber and Romberg tests and what each is useful for measuring.

Vision

Sight is the most dominant of all the human senses. Over 70% of the body's receptors are the specialized photoreceptors found in the retina of the eye. The human eye and its accessory structures provide the structural apparatus required for processing and receiving visual images.

Exercise 1 Anatomy of the eye

Label each of the terms found in bold on the eye diagrams at the end of the chapter. Use charts, the descriptions in the text and dissectible models of the human eye to assist you in the identification.

The **extrinsic eye muscles** are attached to the outer wall of the eyeball. There are 6 of these muscles in all that are responsible for moving the eyeball in the various directions.

The wall of the eye is made up of 3 different tunics or coats.

The **fibrous tunic** is the outermost coat. The sclera makes up the posterior five-sixths of the fibrous tunic and is commonly referred to as "the white of the eye". The sclera serves as an attachment sight for the extrinsic eye muscles. The **cornea**, which makes up the anterior 1/6^{th,} is the window of the eye. Although it is made up of fibers and some cells, it is transparent. The fibrous tunic has a very poor blood supply.

The middle layer of the eye is the **vascular tunic** or uvea. The posterior portion is the thin, heavily pigmented choroid. The pigments in this layer help to absorb excess light. Towards the front of the eye the choroid is modified to form the **ciliary body** and the **iris.** The ciliary body is found at the level of the lens and has muscles associated with ligaments attached to the lens that assist the lens in changing shape. The iris is the colored portion of the eye. It is made up of 2 muscles that control the diameter of the pupil. The **pupil** is not an anatomical structure, but rather the hole found in the center of the iris. Light passes through the iris to deeper portions of the eye.

The inner layer of the eye is known as the **sensory tunic** or retina. The retina extends anteriorly only to the ciliary body. The retina contains the photoreceptors: rods and cones. These begin the chain of electrical events that result in action potentials being sent to the brain as a result of light entering the eye. Vision is the final product. Cones are best for color vision and rods function in black and white vision. The photoreceptor cells are distributed over the entire retina except where the optic nerve leaves the eyeball. This site is called the optic disc, or **blind spot**. At the exact center of the eye there is a special region called the **macula lutea**. This is an area of high cone density. In the center of the macula lutea is the fovea centralis, a minute region about ½ mm in diameter that contains only cones and is the area of greatest visual acuity.

The **lens** divides the eye into an **anterior cavity** and a **posterior cavity**. The anterior cavity is filled with a watery fluid called aqueous humor. The aqueous humor is produced and drained from the anterior cavity regularly. Any abnormal increase in the amount of aqueous humor (because of too much fluid being produced or not enough fluid being drained), results in glaucoma. There are many types of glaucoma, but all types result in an increase in the pressure inside of the eyeball, which, if not corrected, can lead to blindness. Pressure on neurons can result in their death and the loss of vision. The posterior cavity is behind the lens and is filled with a transparent gel known as vitreous humor. The vitreous humor prevents the eyeball from collapsing

Exercise 2 Vision Tests

Location of the blind spot

To determine the blind spot in your left eye, hold the figure containing the cross and dot 18 inches in front of your face. Close your right eye and focus on the circle with your left eye. Slowly bring the page closer, still focusing on the circle. The point at which the cross disappears indicates the blind spot, where the optic nerve enters your left eye. This test will not work if you move your eye back and forth.

To determine the blind spot in your right eye, hold the figure 18 inches in front of your face. This time close your left eye and focus on the cross. Slowly bring the page closer, still focusing on the cross. The point at which the circle disappears indicates the blind spot for the right eye.

●a) Why does the image disappear when you find the blind spot (light is falling on what part of the retina)?
●b) Is the blind spot in the peripheral field of vision (toward the side) or the middle field of vision?

Refraction and the path of light through the eyeball

Refraction, or the bending of light occurs when light passes through the various anatomical parts of the eye. Light rays entering the eye pass through many structures that bend the light. Light first passes through the cornea, then the aqueous humor of the anterior cavity, followed by the lens, and finally the gelatinous vitreous humor of the posterior cavity. Refraction is important in focusing light properly onto the retina. All of the refractive structures of the normal eye remain relatively constant in their light bending capacity except for one...the lens.
The **lens** plays a key role in refracting light more or refracting light less depending on the distance of the object being viewed. The lens is a biconvex disc housed in a thin fibrous bag. Altering the tension produced by the ligaments and muscles surrounding it can

change its shape. In general, the closer the object, the more the light must be refracted to focus it on to the retina. This is done by allowing the lens to become even more bulgy (more biconvex). Distant objects (more than 20 feet) require little accommodation of the lens to produce a proper image on the retina.
In a normal eye light can be focused on the retina properly by the lens. Visual problems result when (1) a lens is too strong or too lazy, (2) the eyeball has a structural default such as being too long or too short, or (3) a cornea has improper curvatures. **Myopia** or "nearsightedness" results when light is focused too soon...before hitting the retina. Nearsighted individuals can see well close up, but distant objects are blurred. **Hyperopia**, or "farsightedness" results when the ideal image would be produced somewhere behind the retina...light rays would converge behind the eyeball. As a result, farsighted individuals have no problem with distant vision, but need glasses with a convex lens to help to bend light soon enough to see near objects well.

Testing for the flexibility of the lens

The lens loses much of its capacity to change shape as it ages. This results in a difficulty viewing very close objects. This condition is called **presbyopia** –literally, old vision. This is why adults frequently hold the newspaper at arms length while reading. The elasticity of the lens can be determined with a test called the "**near point of accommodation**" test. The near point of vision in young adults is about 10cm. It is nearer in children and farther in old age. To determine your near point of accommodation get a ruler and a pin (or you may use the tip of a pencil). Hold the pin at arms length and close one eye while focusing on the pin with the open eye. Slowly move the tip of the pin toward you until the image becomes distorted. Have your lab partner measure this distance; this is your near point of accommodation.

●c) Record the distance for the near point of accommodation for the right eye
●d) Record the distance for the near point of accommodation for the left eye

Test for astigmatism

Irregularities in the curves of the cornea and/or lens lead to a vision problem called astigmatism. View the astigmatism chart without glasses first with one eye, then the other. If all the radiating lines appear equally dark, straight and distinctive, there is no problem with your refractive surfaces of the lens or cornea. If some of the lines appear darker, blurred or wavy you have at least some degree of astigmatism. Astigmatism can be corrected by grinding lenses to compensate or with laser surgery to the cornea.

●e) Is astigmatism present in the right eye?
●f) The left eye?

Visual Acuity

Have another student in the lab stand at the Snellen eye chart on the wall. Read the letters on the chart at a distance of 20 feet. By covering each eye in turn, you can determine the approximate sharpness (acuity) of vision in your right and left eyes. To interpret your results, have your partner read you the numbers that correspond to the last line that you called off that was entirely correct. Normal vision is 20/20, which means that an individual being tested can read letters at a distance of 20 feet that a person with normal vision can see at 20 feet. If vision were 20/40, the person being tested would see at 20 feet what an individual with normal vision could see at a distance of 40 feet.

●g) Record your results for the left eye
●h) Record your results for the right eye
●i) Is your vision above or below average?

Exercise 3.

Microscopic anatomy of the cell layers of the retina

Obtain a slide of the eye. Identify the cell layers of the retina by comparing it to the figure. Notice that as light passes through the eye it must pass through the first two cell layers of the retina before striking the **photoreceptor layer** (rods and cones). The rods and cones then influence the neurons that make up the other two cell layers. The rods and cones target the **bipolar neurons** in the middle of the retina, which then affect the **ganglion cell** neurons. The long processes of the ganglion cells travel together towards the back of the eyeball and exit the eye at the optic disk. These long processes make up the optic nerve, which carries action potentials to the visual centers of the brain.

●a) Draw the 3 cell layers (photoreceptor, bipolar neuron, and ganglion cell layers) of the retina from the slide and label each.

Smell and Taste

The senses of smell and taste are classified as chemical senses because they use chemoreceptors that respond to chemicals of the external environment that are in solution. The mucous lining the nasal cavity and the saliva in the mouth are very important in allowing the receptors to detect the chemicals. The sense of taste includes at least 4 basic types of receptors responsible for the identification of 4 distinct qualities of taste (sweet, sour, salty, bitter). A flavor like chocolate would stimulate both sweet and bitter receptors. Chemical receptors in the nose are responsive to a much wider range of chemical stimuli than those in the mouth.

Location of the olfactory receptor cells

Locate the bold terms in the following text on the diagram and describe the role of each in olfaction. The **olfactory epithelium** is the tissue that includes the olfactory receptor cells. The olfactory epithelium occupies an area of about 2.5 cm in the roof of each nasal cavity. The receptor cells that detect chemicals are assisted in performing their function by other structures in the nasal cavity. There are many mucous secreting glands amid the receptors that produce a watery lining over the receptors. This mucous lining is important in trapping and dissolving particles (odorants) that enter the nose so that they may be detected. There are also small bones in the walls of the nasal cavity called "**turbinates**". The turbinates increase the chances that particles entering the nasal cavity will come in contact with the olfactory epithelium...especially when you sniff.

1. Olfactory Epithelium
2. Turbinates

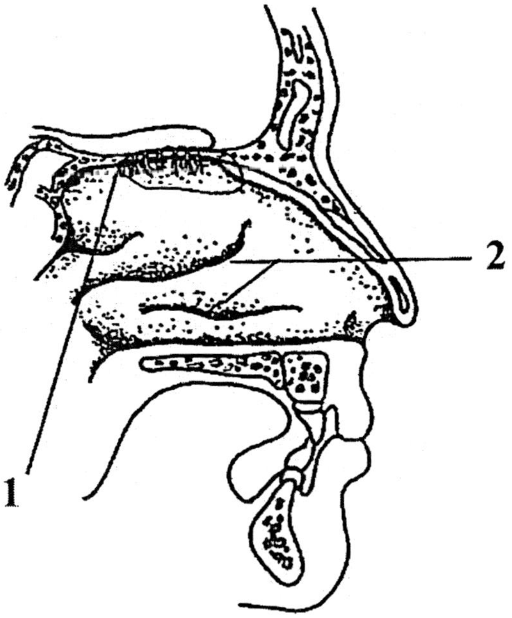

Exercise 4

Tongue anatomy: location of papillae and taste buds

Observe a prepared slide of a cross-section of the tongue. The tongue is covered with 3 different sizes of small mounds called **papillae**. Filiform papillae are very thin spiked projections that give the tongue a velvety appearance. Fungiform papillae or more rounded and scattered across the tongue. Toward the back of the tongue there are very large circumvallate papillae arranged in a V-shape. See how many types you can observe on the tongue of another student. The two larger types of papillae house the special sensory organs for taste, the **taste buds**. On the slide see if you can identify a papilla on scan power. After identifying the papilla, move to high power and identify the onion-shaped taste buds on the wall of the papilla.

●a) Make a drawing of your observations labeling the taste bud and papillae.

Exercise 5. Taste and smell tests

Localization of the various qualities of taste

There are at least 4 different distinct qualities of taste: sweet, sour, salty and bitter. Taste buds scattered across the tongue and mouth generally detect just one of these qualities. Using a cotton-tipped applicator, dip a clean end in the sweet solution. Touch the applicator to your tongue in the following areas: the tip, the front sides, the back sides, just beyond the tip, the middle and the back middle.
●a) Label on the tongue diagram where you could best detect this taste. Repeat the procedure for the other 3 qualities of taste. In most people, sweet and salty are toward the tip, sour is on the sides and bitter in the back.
●b) How do your results compare with the average tongue?

The effect of smell on taste

As you might expect, the sense of taste is very dependent on our other senses. The sense of smell, touch and even pain (hot, spicy foods) play a role in our appreciation for foods. Working with another student, have one person administer the following test while the other is the guinea pig. The person volunteering should stop reading at this point. The person administering the test will dip two cotton applicators in two different concentrated flavor extracts. Return to the table and ask your partner to close his eyes and stick out the tongue. Waft one of the extracts back and forth under the nose while stroking the tongue with the other flavor. Ask your partner what flavor was tasted (you may need to generalize and ask them if it was like vanilla, mint, fruity).

●c) Was your partner able to accurately state the flavor tasted?
●d) Based on these results and your life experience, do you think smell has a large effect on taste?

Hearing

The ear has three parts: the outer, middle and inner ear. For hearing to occur, sound waves must travel into the outer ear, be amplified by the middle ear and converted to an action potential by receptors in the inner ear. The brain must then interpret the action potentials to complete the process of hearing.

Exercise 6. Ear anatomy

Familiarize yourself with the following ear structures and functions below. After studying the available diagrams and models, label the bold terms on the diagram at the end of the exercise.

The **outer ear (external) ear** is comprised of the **auricle** (pinna) and the **external auditory canal**. The pinna is the external ear flap that protects the auditory opening and funnels sound waves toward the auditory canal. The external auditory canal is the inch long passageway leading to the **tympanic membrane** (eardrum). The passageway is lined with ceruminous glands that produce earwax, or cerumen. Once sound waves enter the external auditory canal, they eventually strike the tympanic membrane. This membrane separates the outer from the middle ear.

The **middle ear** is a small air-filled cavity containing the **auditory ossicles**. The auditory ossicles are the tiniest bones in the body. The three bones are the **malleus, incus**, and **stapes**. As sound waves strike the tympanic membrane they cause it to vibrate. These vibrations are then transmitted across the middle ear bones to the **oval window** of the inner ear.
The middle ear also includes the **auditory tube**. The auditory tube links the middle ear to the throat. This tube is normally flattened, but yawning or swallowing can open it to equalize the pressure in the middle ear with the atmospheric pressure. This link to the throat an sometimes allow microbes of the throat access to the middle ear causing the painful middle ear infection called otitis media.

The inner ear consists of snail-shaped **cochlea** and the **vestibular apparatus**. Both structures are fluid filled chambers with receptor cells that respond to fluid movements and create action potentials. The cochlea is involved with hearing and the vestibular apparatus functions in our sense of balance and equilibrium. The vestibular apparatus is made up of the three **semicircular canals** an enlargement adjacent to the cochlea called the **vestibule.**

Exercise 7. Hearing and balance tests

The two hearing tests you will do are often used in a clinical setting to identify patients with suspected hearing impairments. The Rinne test is used to diagnose **conduction impairment**. Conduction impairment is a result of the outer or middle preventing the transmission of sound waves or vibrations. Too much earwax would be an example of a cause for this type of hearing deficit. The Weber test is used to determine conduction deafness or possible nerve impairment. **Nerve impairment** is a result of nerve damage in the inner ear or in the pathway from the inner ear to the brain.

Conducting the Rinne Test:

Strike a tuning fork on the palm of your hand. Place the handle against your partner's head on the bump just behind the ear. When your partner no longer hears the hum of the tuning fork, quickly move the prongs to the opening of the external auditory canal (be sure not to touch the ear). Normally, a hum will be heard again. If not, a conduction impairment is suspected. Repeat the procedure on the opposite ear.

●a) What type of hearing deficit does the Rinne test detect?

●b) Did your partner have an impairment?

Conducting the Weber Test:

Strike the tuning fork with the palm and place the handle in the middle of the forehead. Ask your partner if the sound seemed louder on one side than another. If the sound seemed louder on one side; that may mean that the ear that it appeared louder in has a degree of conduction deafness or the opposite ear may have nerve impairment. Perform the test again, this time simulate a conduction impairment by plugging one ear.

●c) Was the sound heard better in the right ear, left ear or was it equal to both ears?
●d) When you plugged one ear, was the sound louder or softer in the plugged ear (do the test, it may not be what you suspect)?

Conducting the Romberg Test:

The Romberg test demonstrates the ability of the subject to maintain balance and equilibrium using the information from the vestibular apparatus without the help of vision. Have your partner stand near the chalkboard, facing away from the board. Use the desk lamp to cast a shadow of your partner onto the board. With feet together and eyes open, have the subject stand 2 minutes. Mark the farthest edges on the board as the subject sways. After 2 minutes measure the distance of the longest sway.
Repeat the procedure with the eyes closed.

●e) Was the length of sway greater with the eye open or closed?
●f) What other sense works together with the middle ear to maintain equilibrium

The Special Senses

Name ______________
Section ______________

Exercise 1

Label the eye diagrams with the numbered terms provided

1. Cornea (2x)
2. Pupil (2x)
3. Iris (2x)
4. Extrinsic Eye Muscles
5. Anterior Cavity w/Aqueous Humor
6. Fibrous Tunic
7. Vascular Tunic
8. Sensory Tunic
9. Macula Lutea
10. Posterior Cavity w/Vitreous Humor
11. Lens

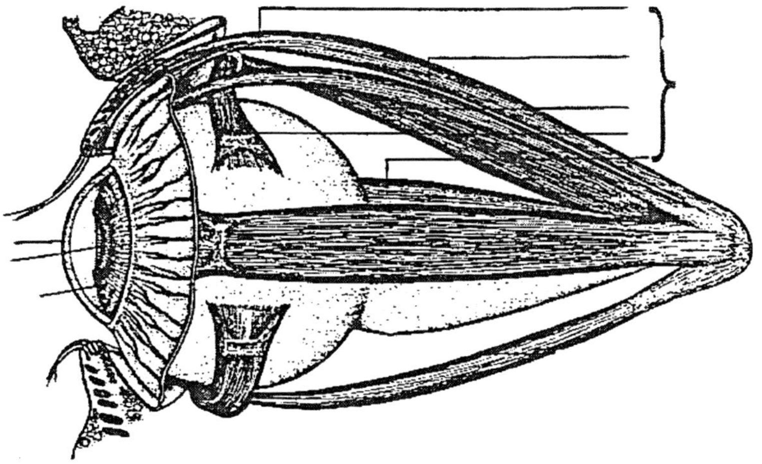

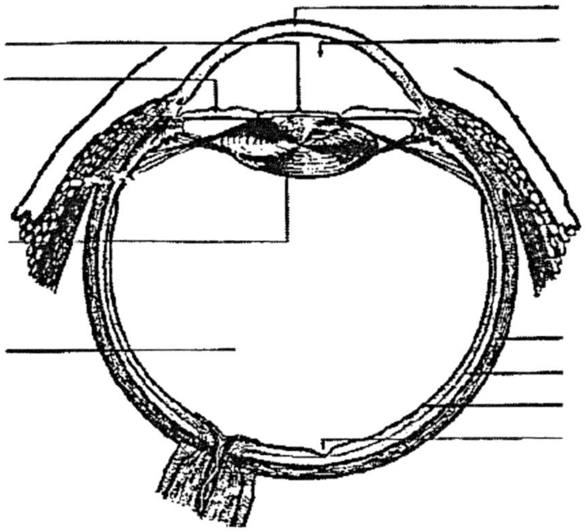

Exercise 2.

a. ____________________
b. ____________________
c. ____________________
d. ____________________
e. ____________________
f. ____________________
g. ____________________
h. ____________________
i. ____________________
j. ____________________

Exercise 3.

Draw a section of the retina and label the ganglion cell layer, the bipolar neurons layer and the rods and cones layer.

Exercise 4.

Draw a papilla and label the taste bud

Exercise 5.

a. Diagram a U-shaped tongue and label where you detected sweet, salty, sour and bitter on the tongue below

b. ____________________
c. ____________________
d. ____________________

Exercise 6. Label the ear diagram below using the following terms: outer ear, auricle, external auditory canal, tympanic membrane, middle ear, inner ear, semicircular canals, and cochlea.

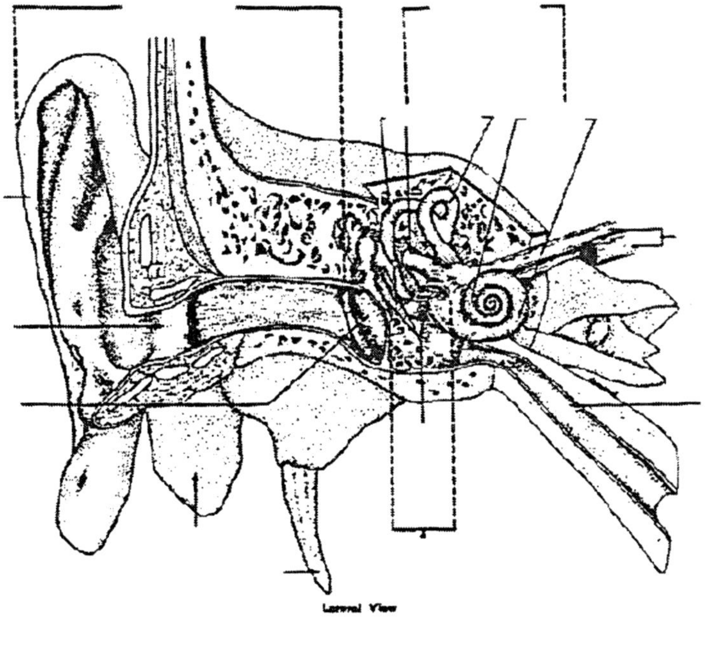

Exercise 7

a. ____________________
b. ____________________
c. ____________________
d. ____________________
e. ____________________
f. ____________________

Animal Behavior

Introduction

An animal's **behavior** can be thought of as its response to stimuli in the environment, or more simply, what an animal does. Thus, behavior can range from predator avoidance, territorial defense, courtship displays, or hunting/foraging for food, to an animal's sleep habits and patterns. Behaviors are often divided into two categories. Those that are **innate** or **instinctive** are said to be genetically programmed, and can be performed by the animal without previous experience or practice. For example, dogs naturally bark at unknown or unusual stimuli without being taught to do so. Their barking can be thought of as being an innate behavior. Those behaviors that are **learned**, on the other hand, can be modified over time by the animal's previous experience. For example, both dogs and cats readily learn to associate the sound of an electric can opener or similar stimulus at a certain time of day with feeding time, and usually do not need an invitation to come. While the innate/learned behavior distinction is a useful one, it is incorrect to classify a behavior as being due to either genes only or to experience only. This was the false division made during the "**nature versus nurture**" debate, when each behavior was classified as being due to either an animal's genetics only (nature), or to an animal's environment and learning only (nurture). We now appreciate that every behavior can be thought of as being determined by a combination of an animal's genes working in conjunction with the environment and experiences. Dogs bark instinctively, but they are capable of learning to discriminate between strangers and friends, and to bark only at strangers. Likewise, a learned behavior never occurs in isolation from the genes that ultimately underlie that behavior. Many birds learn the song that they will sing as adults. However, most of these same birds cannot learn to sing all songs imaginable. Rather, they are genetically programmed or "hardwired" to learn to sing a certain type of song that is characteristic of its species. In today's lab we will observe the territorial display of the common aquarium pet, the Siamese fighting fish, *Betta splendens*, and consider how this behavior is ultimately designed to ensure the survival of this species.

Objectives

Upon completing your study of this unit, you should be able to:

1. Define behavior, innate behavior, and learning.
2. Summarize the "nature versus nurture" debate, and explain why it is incorrect to say that a behavior is due either to an organism's genes only (nature) or to its previous experience only (i.e. nurture/learning).
3. Define display and communication, using the concepts of a sender and a receiver.
4. Explain the concepts of stereotyped behavior, fixed action pattern, and sign stimulus in terms of innate behaviors.
5. Offer a hypothesis as to whether color, shape (appropriate/exact or abstract), or both are important sign stimuli in triggering the aggressive displays of male B*ettas*, based on your lab observations.
6. Define territoriality and dominance hierarchy, and explain how these types of aggressive behaviors can be advantageous to individuals.

Exercise 1.

Place a container holding a single male *Betta* in the middle of the table where it can be observed by everyone in your group without disturbing the fish. Observe the fish's normal resting or submissive posture and color for five minutes. ●a) Make a sketch of the fish's general body shape, including the shape of the fins. Compare your sketch with the illustrations below that show both the resting/submissive posture and the aggressive broadside and frontal postures. ●b) Which does your sketch more closely resemble? As you are observing your fish's response to no stimulus (a control situation), record on the data sheet with a check any behaviors observed. After the five minutes of observation also record with a check one of the four overall responses to this control situation.

When displaying aggressively, as you will hopefully see shortly, a *Betta* will first exhibit a "broadside display" in which it turns sideways to an intruder and fans its fins out. If aggression mounts, the displaying male will eventually face the intruder in the frontal display, raising its gill covers. A fish's color will also often darken with increased aggression. If the intruder does not withdraw, a frontal attack may occur. ●c) Why do you think a male *Betta* turns sideways and fans its fins out; how is it "trying to look"? ●d) How is a *Betta* "trying to look" when it pulls its fins in, and what message do you think is being sent by this posture? A male *Betta* will also assume the broadside display when it is courting a female *Betta* that it wishes to attract to its nest to lay eggs. ●e) Which type of display (submissive or aggressive) would you predict a female *Betta* needs to give for successful nesting to occur?

A display, such as the broadside display that we observe in the *Betta*, is an example of a behavior with the function of communication between two individuals. **Displays** are behaviors in which the action(s) is exaggerated and/or stereotyped in order to get the message across. The display is said to serve as **communication** because one individual is functioning as a **sender** of information, and is transmitting a signal (in this case a nonverbal one) to a **receiver**. In the case of communication between two male *Bettas* using the broadside display, the signal is meant to be a threat and to drive away a competitor. In the case of male-female communication between *Bettas* using this display, on the other hand, the signal is meant to impress and attract the female.

Resting/Submissive Display

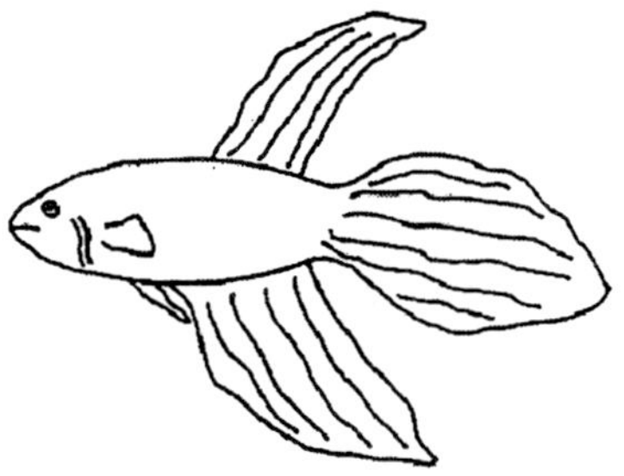

Broadside Display

Frontal Display

Exercise 2.

You will now attempt to stimulate your fish to give the broadside display. Take the mirror provided and position it so that your male can see its own image at close range. It may take a moment for the fish to notice the "intruding male" and to respond. As it does, compare its posturing with the illustration in your lab book, and also record on your data sheet with a check in the appropriate column any behaviors observed. ●a) Does its broadside display look similar to that illustrated, or does it differ quite a bit? The broadside display of the *Betta* is an example of a behavior that is **stereotyped**, i.e. it is normally given by individuals in a nearly identical manner. ●b) Do you think such a stereotyped behavior should be considered primarily innate or learned? To check your answer, consider the fact that experiments show that *Bettas* can give a broadside display even if the fish has never given it before nor seen it performed previously.

Such a rigidly given innate behavior or display is referred to as a **fixed action pattern** (FAP). An FAP, after being triggered by the appropriate **sign stimulus**, will automatically be performed by the individual to completion and with minimal variation in its details, almost like a machine. Another example would be the arched back and raised fur of a cat when it is frightened. A sign stimulus that triggers an FAP is typically a simple signal with only one or two important components. In Exercise 3 we will attempt to determine the crucial components of the sign stimulus that triggers the broadside display in *Bettas*.

Exercise 3.

In your materials are six models that will be used to try and determine the important characteristics of the sign stimulus that triggers the innate broadside display in *Bettas*. Below is a summary of each model's characteristics. They will help us determine if both color and shape are important aspects of the sign stimulus, or if only one of these is important. If shape is important, we will try to decide if the shape needs to be exact (appropriate), or if a general abstract shape is a sufficient sign stimulus.

Model characteristics:

Model A Inappropriate shape (vertical oval), inappropriate color (white)
Model B Abstract shape (triangle), inappropriate color (white)
Model C Appropriate/exact shape, inappropriate color (white)
Model D Inappropriate shape (vertical oval), appropriate color
Model E Abstract shape (triangle), appropriate color
Model F Appropriate/exact shape, appropriate color

You will present these six models to your male B*etta* one at a time in the order listed, beginning with model A and ending with model F. Each time a model is presented hold it up to the outside of the fish jar so that your fish can see it (Do not put the models into the water with the fish). Move slowly so as to not alarm your fish. Present the model to your fish for a total of three minutes, sometimes holding it motionless, and sometimes moving it back and forth at a slow swimming speed. As each of the six models is presented, have a lab member record the male B*etta's* reaction to that model on the data sheet, putting a check for each type of reaction observed during the three minute test period. After the fish has been tested for one model for three minutes, remove the model and record what the fish's overall response to that model was during the three minutes on the data sheet. Between each model's presentation wait one or two minutes before presenting the next model, in order to give the *Betta* time to recover from the previous presentation.

Exercise 4.

As a final exercise your lab instructor will place two male B*ettas* together into the same container so that the class can observe a male's response to another live male B*etta*. As the fish interact, again record on your data sheet in the appropriate column any reactions of either male to the other. In the most aggressive encounters involving a frontal attack one male will attempt to bite and tear the fins of another male. The fish will be separated if the aggression becomes too intense, in order to prevent serious injury to either fish.

Aggressive displays are a common type of social behavior among many species of animals. Often the display serves to advertise an individual's **territory**, which is an area defended from other individuals. For example, early in the spring male robins may fight to establish territorial bounds. After this, however, the singing of a male robin on its territory is usually enough to warn other male robins away. Fighting to establish a territory can be a risky behavior because of the possibility of injury. It also costs the individual energy. ●a) What might the benefit be to an individual in establishing a territory that would outweigh these disadvantages?

Aggressive displays also sometimes serve to establish a **dominance hierarchy**, which is a social ranking in a group of individuals of the same species where more submissive or subordinate members give way to more dominant group members. For example, within a wolf pack a new individual joining the pack has to interact with other pack members and usually has to engage in some fighting in order to establish its position in the pack's dominance hierarchy. Again, the initial fighting is usually a risky or expensive behavior. However, once an animal has established its rank within a dominance hierarchy, it does not need to fight again and again with the same individuals constantly. Instead, subtle gestures are sufficient to maintain order. Although an individual low in the dominance hierarchy may have to wait a while before it gets access to a resource such as food, sooner or later it does usually get access. It just may not get the best pickings. ●b) Considering these facts, what benefit is there to each individual in the pack in establishing its place in the dominance hierarchy?

●c) Relating these ideas back now to the B*etta* displays that we have been observing, why is there an advantage for these fish to have a threatening display that can be used to determine who is stronger, rather than just fighting with every male each time they meet? ●d) These Siamese fighting fish have been bred in captivity to have a more aggressive nature than is typical in the wild. If such an individual was released back into the wild in the streams of Indonesia, do you think it would be at an advantage or disadvantage to have this artificially high aggressive nature? Think about what that individual's likely fate would be if its fighting behavior was triggered by every male that challenged it, even if that challenger was bigger and stronger than itself.

Name ______________________________

Number______________________________

Animal Behavior

Exercise 1

a.

b. ______________________________________

First record any behaviors observed by the male B*etta* in the appropriate column of the data table below when in a control (i.e. no stimulus) situation

			Model						
Reaction of fish	**Control**	**Mirror**	**A**	**B**	**C**	**D**	**E**	**F**	**Fish**
No response									
Swimming									
Swimming away									
Spreading of top (dorsal) fin									
Spreading of bottom (ventral) fin									
Spreading of tail (caudal) fin									
Darkened coloration									
Raising of gill covers									
Facing stimulus									
Biting at stimulus									
Overall Response									
No response									
Retreated									
Broadside Display									
Frontal Display/Attack									

c. ______________________________

____

d. ______________________________

____

e. ____________________

Exercise 2

Record in the appropriate column of the data table any behaviors observed by the male B*etta* in response to its mirror image.

a. ____________________

b. ____________________

Exercise 3

Record in the appropriate columns of the data table any behaviors observed by the male B*etta* in response to each of the models presented one at a time.

Exercise 4

Record in the appropriate column of the data table any behaviors observed by either of two male B*ettas* when placed together in the same container with each other.

a. ______________________________

____

____

b. ______________________________

____

____

c. ______________________________

____

____

d. ______________________________

____

____